THE ECOLOGY OF BRUCHIDS ATTACKING LEGUMES (PULSES)

SERIES ENTOMOLOGICA

EDITORS

E. SCHIMITSCHEK & K. A. SPENCER

VOLUME 19

SPRINGER-SCIENCE+BUSINESS MEDIA, B.V. 1981

THE ECOLOGY OF BRUCHIDS ATTACKING LEGUMES (PULSES)

Proceedings of the International Symposium held at Tours (France), April 16-19, 1980

Edited by
V. LABEYRIE

SPRINGER-SCIENCE+BUSINESS MEDIA, B.V. 1981

Symposium organized by I.B.E.A.S. (Université Francois Rabelais and C.N.R.S.: French National Centre of Scientific Research) in collaboration with the I.O.B.C. and the F.A.O.

This volume is listed in the Library of Congress in Publication data

ISBN 978-90-481-8525-2 ISBN 978-94-017-3286-4 (eBook)
DOI 10.1007/978-94-017-3286-4

Cover design: Max Velthuijs

Originally published by Dr W. Junk Publishers, The Hague in 1981
Softcover reprint of the hardcover 1st edition 1981

PREFACE

As editor of this volume I am not going to emphasize the economic importance of the Bruchids of pulses, nor how necessary it is to know the conditions of infestation of crops in order to achieve an efficient protection; both points are testified by F.A.O.'s sponsorship, as well as of the International Organization for Biological Control. On the other hand I would like to stress the scientific interest of the study of 'domestic' bruchids. It raises questions which require that present basic entomological knowledge be extended. I am not going to review all questions raised but I will emphasize those I am most aware of.

Some bruchid species have been able to colonize habitats differing totally; some differing in their latitude, and thus their basic periodicities; others differing in their degree of complexity (stocks of seeds):

What are the respective roles played by polymorphism and plasticity in that exceptional capacity?

What differences are there between the populations living on wild plants in dispersal areas and those living in stocks?

What are the alterations brought about by the drastic selection pressures in populations having later colonized habitats ecologically similar to the original habitats?

What factors determine the degree of specificity of trophic relationships, and the possibilities of extension of the niche?

What influences are exerted by the other elements of the original biocenoses upon the behavioural evolution of these species?

Such are a few fundamental problems that can be tackled directly through bruchids. Any progress in our knowledge of these phenomena will have immediate repercussions on the improvement of economic methods for the prevention of attacks and the protection of crops whose destruction seriously impairs the protein food of poorer populations.

But these questions can be adequately answered only by combined research involving biochemists, physiologists and ethologists in partnership. The qualitative and quantitative evolution of bruchid populations, and their coevolution with legumes, cannot be examined without such cooperation transcending botany and entomology.

Thus, the methodology adopted to solve the human problems generated by bruchid-caused destructions will provide a model for an approach to crop protection that will be really ecological, being systemic.

This orientation will not necessitate a new terminology, but it will require giving up watertight administrative compartments in the science of nature, which has been cut into slices by positivist scholasticism.

The recommendations to F.A.O., and the creation of a working group attached to the I.O.B.C., including botanists and ethnologists as well as entomologists, show that applied entomologists are not doomed to carry on their solitary work.

Thus in addition to what will be added by the various papers to the knowledge of the bruchids' ecology, this colloquium has already achieved an important result that should be remembered.

The manifold scientific activity of Professor G.S. Fraenkel, Chairman of the colloquium, should be a lesson to us. It is by reaching across the boundaries of botany and entomology, that he could show the importance of allelochemical substances in the coevolution between plants and consumers. This work is dedicated to him, on the occasion of his eightieth birthday, to the new Doctor Honoris Causa of the University, which is proud of bearing the name of the great humanist of the French Renaissance, François Rabelais, the discoverer of pheromones.

Tours, August 1980 Vincent Labeyrie

CONTENTS

The participants of the Symposium at a party organized by the Country Council (Conseil Général) at the priory 'Prieuré de St. Cosme', where lived the great Renaissance poet, Pierre de Ronsard.

1. C.D. Johnson
2. M. Hodkovà
3. M. El Agoze
4. Représentant du Préfet
5. G.S. Fraenkel
6. L.E. Ramirez
7. G. Mathys
8. V. Labeyrie
9. M. Moreire
10. M. Dakkouni
11. B.J. Southgate
12. D. Thiery
13. I. Alzouma
14. D. Stamopoulos
15. R. Camut
16. J.P. Katz
17. M. Sidibe
18. Mrs. Alzouma
19. P. Anglade
20. B. Leroi
21. A. Piavaux
22. J. Brenière
23. D. Pierre
24. R. Kalmès
25. C. Cafati
26. J. Huignard
27. J.C. Biemont
28. J.C. Landré
29. Y. Lozano
30. P. Garaud
31. A. Garcia Calleja
32. J.P. Cancela da Fonseca
33. P. Lagrange
34. P. Dobie
35. A. Lawson
36. M. Jarry
37. J. Decelle
38. A. Ouedraogo
39. J. Pouzat
40. S. Utida
41. G.A. Rosenthal
42. C. Thibeaudeau
43. M.H. Hamelin
44. B. Diallo
45. A. Chacon
46. A. N'Diaye
47. 1er interprète
48. 2ème interprète
49. A. Bonet
50. P. Desroches
51. H.R. Pajni
52. C. Arnault
53. J. Jones
54. M. Pimbert
55. H. Dell Orto
56. G. Hariri

LIST OF PARTICIPANTS

Abo Ghalia, M.	Ain Sham University Department of Entomology, Cairo, Abbassia (Egypt)
el-Agoze, M.	Ain Shams University, Department of Entomology, Cairo, Abbassia (Egypt)
Al Moussa, N.	University of Aleppo (Syria)
Alzouma, I.	Biological Laboratory, University, Niamey (Niger)
Anglade, P.	Station de Zoologie Agricole, Centre de Recherches Agronomiques du Sud-Ouest, I.N.R.A., La Grande Ferrade, 33140 Pont de la Maye (France)
Arnault, C.	I.B.E.A.S., Université F. Rabelais, Avenue Monge, Parc Grandmont, 37200 Tours (France)
Biemont, J.C.	I.B.E.A.S., Université F. Rabelais, Avenue Monge, Parc Grandmont, 37200 Tours (France)
Bonet, A.	Instituto de Ecologia, Nuevo Bosque de Chapultepec, Ap. postal 18-845, MEXICO 18 D.F. (Mexico)
Boscher, J.	I.B.E.A.S., Université F. Rabelais, Avenue Monge, Parc Grandmont, 37200 Tours (France)
Botte, F.	I.B.E.A., Université F. Rabelais, Avenue Monge, Parc Grandmont, 37200 Tours (France)
Brader, L.	Plant Protection Service (Plant Production and Protection Division), F.A.O., Via Delle Terme di Caracalla, 00100 Rome (Italy)
Breniere, J.	I.R.A.T./G.E.R.D.A.T., B.P. 5035, 34032 Montpellier CEDEX (France)
Buess, H.U.	Sandoz S.A. Agrodivision, 4002 Basel (Switzerland)
Cafati, C.	Instituto de Investigaciones Agropecuarias, Estacion Experimental La Platina, Casilla 5427, Santiago (Chile)
Camut, R.	VILMORIN, La Ménitré, 49250 Beaufort en Vallée (France)
Cancela da Fonseca, J.P.	Laboratoire de Biologie Végétale, Route de la Tour Dennecourt, 77300 Fontainebleau (France)

Cepede, M.	Comité Français de l'Organisation des Nations Unies pour l'Alimentation et l'Agriculture, 30 rue Las Cases, 75007 Paris (France)
Chacon, A.	Escuela de Biologia, Universidad de Costa Rica, San José (Costa Rica)
Chamoux, M.	Procida, 103 Avenue Gambetta, 75020 Paris (France)
Combescot, C.	I.B.E.A.S., Université F. Rabelais, Avenue Monge, Parc Grandmont, 37200 Tours (France)
Dakkouni, M.	Faculté des Sciences, Rabat (Morocco)
Decelle, J.	Musée Royal de l'Afrique Centrale, Departement d'Entomologie, Tervuren (Belgium)
Dell'orto, H.	Instituto de Investigaciones Agropecuarias, Estacion Experimental La Platina, Casilla 5427, Santiago (Chile)
Desroches, P.	I.B.E.A.S., Université F. Rabelais, Avenue Monge, Parc Grandmont, 37200 Tours (France)
Diallo, B.	Institut de Technologie Alimentaire, Route des Pères Maristes, B.P. 2765, Dakar (Sénégal)
Dobie, P.	Tropical Products Institute, Slough (England)
Errard, C.	I.B.E.A.S., Université F. Rabelais, Avenue Monge, Parc Grandmont, 37200 Tours (France)
Fraenkel, G.S.	University of Illinois, Department of Entomology, Urbana (U.S.A.)
Garaud, P.	I.B.E.A.S., Université F. Rabelais, Avenue Monge, Parc Grandmont, 37200 Tours (France)
Garcia Calleja, A.	Servicio de Defensa contra Plagas e Inspeccion Fitopatologica, c/o Santuario 14, Valladolid (Spain)
Georgel, P.	Université Paris Sud, Laboratoire d'Entomologie, 91405 Orsay CEDEX (France)
Gillier, P.	Institut de Recherches pour les Huiles et Oléagineux (I.R.H.O.), 11 square Pétrarque, 75016 Paris (France)
Gomez, L.	Universidad de Medellin (Colombia)
Hariri, G.	Université d'Aleppo, Faculté d'Agriculture, Aleppo (Syria)
Hecq, J.	Commission des Communantés Européenne, Division Développement, Brussels (Belgium)
Hodek, I.	Academy of Sciences, Prague 2, Vinicna 7, 128.00 (Czechoslovakia)
Hodkova, M.	Academy of Sciences, Prague 2, Vinicna 7, 128.00 (Czechoslovakia)
Huignard, J.	I.B.E.A.S., Université F. Rabelais, Avenue Monge, Parc Grandmont, 37200 Tours (France)
Jarry, M.	C.N.R.S. – I.B.E.A.S., Université F. Rabelais, Avenue Monge, Parc Grandmont, 37200 Tours (France)

Johnson, C.D.	Northern Arizona University, Department of Biological Sciences, Flagstaff (U.S.A.)
Jones, J.	Agricapital Group, c/o B.S.S.R.S., 9 Poland Street, London W1 (United Kingdom)
Katz, J.P.	SEPANT et Société des Sciences Naturelles de Touraine 1 allée Pelletier d'Oisy, 37000 Tours (France)
Kalmes, R.	I.B.E.A.S., Université F. Rabelais, Avenue, Monge, Parc Grandmont, 37200 Tours (France)
Labeyrie, V.	I.B.E.A.S., Université F. Rabelais, Avenue Monge, Parc Grandmont, 37200 Tours (France)
Langley-Danysz, P.	Revue Sciences et Avenir, 68 rue Jean Jacques Rousseau 75001 Paris (France)
Lawson, A.	Ecole Supérieure d'Agronomie, Université du Bénin, B.P. 1515, Lomé (Togo)
Lecomte, C.	I.B.E.A.S., Université F. Rabelais, Avenue Monge, Parc Grandmont, 37200 Tours (France)
Leroi, B.	I.B.E.A.S., Université F. Rabelais, Avenue Monge, Parc Grandmont, 37200 Tours (France)
Lozano, Y.	G.E.R.D.A.T., B.P. 5035, 34032 Montpellier (France)
Massoud, Z.	Laboratoire d'Ecologie du Muséum d'Histoire Naturelle, 4 Avenue du Petit Château 91800 Brunoy (France)
Mathys, G.	Secretary General (I.O.B.C.), 1 rue Lenôtre, 75016 Paris (France)
de Metz Noblat, G.	Grandes Minoteries à Fèves de France, Rue Henri Bessemer, Z.I. Aix les Milles, 13290 Les Milles (France)
Moreira, M.	D.G.P.P.A., Tapada da Ajuda, Lisbon (Portugal)
Morere, J.L.	Université Paris Sud, Laboratoire d'Entomologie, 91405 Orsay CEDEX (France)
N'diaye, A.	Student of Institut Agronomique, 16 rue Claude Bernard, 75231 Paris CEDEX 05 (France)
Ndoye, M.	I.S.R.A. – C.N.R.A., Bambey (Sénégal)
Ouedraogo, A.	Institute Supérieur Polytechnique, Ouagadougou (Upper Volta)
Pajni, H.R.	Panjab University, Zoology Department, Chandigarh (India)
Palauqui, M.	I.B.E.A.S., Université F. Rabelais, Avenue Monge, Parc Grandmont, 37200 Tours (France)
Piavaux, A.	Commission des Communautés Européennes, Rue de la Loi 86, Bureau 4/49, 1040 Brussels (Belgium)
Pierre, D.	I.B.E.A.S., Université F. Rabelais, Avenue Monge, Parc Grandmont, 37200 Tours (France)
Pimbert, M.	University of Liverpool (United Kingdom)
Pollet, A.	O.R.S.T.O.M., 24 rue Bayard, 75008 Paris (France)

Pouzat, J.	I.B.E.A.S., Université F. Rabelais, Avenue Monge, Parc Grandmont, 37200 Tours (France)
Ramirez, L.E.	C.I.A.T., Cali (Colombia)
Renoux, J.	Association des Universités partiellement ou entièrement de langue française (A.U.P.E.L.F.), 173 Boulevard Saint Germain, 75006 Paris (France)
Robert, P.	I.B.E.A.S., Université F. Rabelais, Avenue Monge, Parc Grandmont, 37200 Tours (France)
Rosenthal, G.A.	University of Kentucky, School of Biological Sciences, Lexington (U.S.A.)
Rouse Rojas, D.	C.N.R.S. – I.B.E.A.S., Université F. Rabelais, Avenue Monge, Parc Grandmont, 37200 Tours (France)
Sidibe, M.	S.R.C.V.O., B.P. 438, Bamako (Mali)
Southgate, B.J.	Slough Laboratory, Ministry of Agriculture, Slough (England)
Stamopoulos, D.	Laboratory of Applied Zoology & Parasitology Faculty of Agriculture, Thessaloniki (Greece)
Steffan, J.R.	Muséum National d'Histoire Naturelle, Laboratoire d'Entomologie, 75005 Paris (France)
Taylor, T.A.	Institute of Agricultural Research, University of Ife, Ibadan (Nigeria)
Thibout, E.	C.N.R.S. – I.B.E.A.S., Université F. Rabelais, Avenue Monge, Parc Grandmont, 37200 Tours (France)
Thierry, D.	I.B.E.A.S., Université F. Rabelais, Avenue Monge, Parc Grandmont, 37200 Tours (France)
Utida, S.	Entomological Laboratory, College of Agriculture, Kyoto University, Kyoto (Japan)
Varaigne Labeyrie, C.	Institut Supérieur Polytechnique, Ouagadougou (Upper Volta) I.B.E.A.S., Université F. Rabelais, Tours (France)

ECOLOGICAL PROBLEMS ARISING FROM WEEVIL INFESTATION OF FOOD LEGUMES

V. LABEYRIE
(*Tours, France*)

INTRODUCTION

Protein deficiency affects 500 million people at least (Poleman, 1975), according to FAO's estimates. The will to substantial increase and improvement of food production should be clearly expressed and become an actual concern. Now 'pulse' productions do not appear in international statistics. The C.N.U.C.E.D. (United Nation Conference for Commerce and Development) has not even listed one Legume among the twelve essential farm products of which world reserves should be provided for. Yet, for tens of millions of people, the daily consumption is above 100 g. 'Yet it is the family Leguminosae that shows most promise for producing the vastly increased supplies of vegetable protein that the world will need in the near future. In developing countries especially, cultivation of Legumes is the best and quickest way to augment the production of food proteins' (*Tropical legumes*, 1979).

Through various investigations, we have managed to record 89 legume species whose fruits are consumed by man. A more thorough investigation would perhaps show a hundred of them. But Smartt's work, *Tropical pulses* (Smartt, 1976) only mentions 29 cultivated species, and only 14 of them appear in his bibliographical references. A brief analysis shows that 85% of research works are concentrated on 3 legumes, and half of them on soy bean, though it is not often cultivated for its protein supply to human food.

A few direct investigations, in village shops, at local merchants' and mainly on markets and in peasants' cabins in Columbia as well as in Mexico, High-Volta, Syria or in Guadeloupe, reveal that 80 to 100% grains of *Phaseolus vulgaris*, *Vigna unguiculata* or *Cajanus cajan* are attacked by weevils. But there is not one statistic bearing on these losses; yet preserving these crops would result in a considerable increase in these legume yields.

An efficient protection of crops cannot be expected, by recommending to poor farmers, or to the rural populations of the third-world in general, chemical treatments that involve extra costs, special installations and minimum sanitary equipment to avoid poisoning. Besides, even when the crops have been particularly abundant and part of them goes to storage facilities, the treated crop is none the less depreciated and rejected by the canneries of industrialized countries.

Series Entomologica, Vol. 19: ed. by V. Labeyrie

Everything shows that prevention alone, by preventing crop infestation, can meet these economic requirements, and actually contribute to the improvement of rural or suburban populations' food.

Therefore, it is not paradoxical to study the problems resulting from weevil attacks on food legumes, even though the administrative devision of agricultural entomology departments makes it seem so. In effect, research and application departments are dichotomoized: there is on the one hand the department for the protection of crops dealing with the insects harmful to crops before harvesting, and on the other hand the storage department dealing with the insects that cause losses in stores.

It is evident that this anthropomorphic conception of entomology has allowed efficient curative methods to be devised whenever the storage conditions were intended to favour chemical control.

But this strategy, aiming at successful control of insects ignores all the adjusted techniques aiming at reducing their aggressiveness. Actually, as was mentioned by Whitehead (1930), about bruchids, the failures are due to wrong principles involved rather than to poor application.

These principles are generally rudimentary and compartmental. They are founded only on minimal biological studies, so as to keep research expenses from increasing the cost of intervention. In fact, what is expected from biology is very simple, as was recalled by Geier about ten years ago (1966): 'at the practical level, it' – the contribution of biology to pest control – 'provides the knowledge of biological facts required to devise efficient repressive techniques'.

This conception considers functions, behaviour, life sequences as independent phenomena. Now adaptation is an inclusive phenomenon. There is not one selection pressure operating *alone*, so that each function works in conditions which are never optimal, but result from a *compromise* allowing the other functions to be carried out. An organism is not a 'bean bag', as was phrased by Mayr (1972). Therefore it is impossible a priori to list the aspects of an insect's biology we should know in order to devise an efficient protection.

By attempting to save in this way, applied entomology has sterilized thousands of studies and revealed how vain a number of hopes were. Legume insect pests provide a striking example.

Levins (1973) considers that this fundamental mistake is made even worse in the U.S.A., because within agricultural science, America's traditional anti-intellectualism survives as an antitheoretical bias. Unfortunately the phenomenon is more widespread, as shown by the fact that Henri Fabre – called by Darwin 'the inimitable observer' – wrote in his *Entomological memories*: 'For the awful utilitarian, a quarter of peas preserved from weevils is more important than a number of investigations which are not immediately profitable' (Fabre, 1879).

A quotation from S.R. Singh and H.D. Van Emden's study 'Insect pests of grain legume', published by the *Annual Review of Entomology* in 1977, illustrates these mistakes: 'grain legumes are heavily attacked in storage, particularly by bruchid beetles and some Lepidoptera. Storage pests fall outside the scope of this review, even though the infestation frequently may begin in the field' (Singh *et al.*, 1979).

This significant quotation bears out the fact that the administrative framework can, by opposing biological facts, divide nature into slices, and thus cripple any serious investigation of the infestation process, as if the latter might not *always* begin in the field! Yet it is three centuries since Francesco Redi showed that maggots were not born of meat fermentation!

There are no such things, there can be no such things as stored product insects, but there are insects which can invade and infest stored products. It is because the official trend overlooks this fact that research programs are being fractioned to go into the bed of Procustes.

Thus an entomologist, however experienced he may be (Utida, 1974) was able to write that *Zabrotes subfasciatus* lived only in stores and was a typical example of quite a group of bruchid species showing such behaviour.

Decelle (1972) rightly remarks that the concept of storage insect pest paradoxically makes researchers call 'a normal stage' the one that is characteristic of *Callosobruchus maculatus* living in stores, though it is an abnormal environment. In fact, grain storage cannot induce polymorphism, but on the other hand it must favour some phenotypes in the population.

The spirit in which Mitchell (1975) has studied *C. maculatus*' oviposition tactics in stored grains is even more surprising. He hints, on the one hand, that this insect is able to make plans by divising tactics and, on the other hand that these tactics were divised ever since mankind began storing grains. Actually, the real problem consists in understanding why the behaviour of certain species allows them to oviposit in stored grains.

Since there are no such insects as storage bruchids, we should study how to prevent the infestation of crops by insects which can develop later in stores.

THE BIOLOGY OF BRUCHIDS SHOULD BE STUDIED FROM AN ECOLOGICAL POINT OF VIEW

Preventing crop infestation without knowing the biology of bruchids is obviously impossible. Now what we know in that field is rudimentary and often inaccurate. Southgate (1979) stresses it: 'too little is known about when these beetles lay eggs and how they choose the site'.

Therefore entomological efforts should be reoriented, starting from hypotheses of research aiming at controlling the activity of bruchid populations in order to protect crops.

A clocklike movement as shown by many instances occurring in the history of scientific fashions should not lead to the neglect of problems with direct economic consequences, in order to explore without discrimination the whole range of bruchid biology. This conception of a mass research with 'sheet bombing' relies on chance discoveries to improve knowledge. While the biology of a species makes up such a whole that any function cannot be studied separately, it is true that some aspects should have priority. Choosing these studies depends on the problem to be solved; the hypotheses direct the progress of work.

'Random' research is mystifying. Whoever works without an hypothesis

can hardly help knowledge to progress. Research has nothing to do with digging for nuggets. The facts are not given in their elaborate aspects, they must be brought out of their gangue with the help of specific tools provided by the hypothesis. 'A general view is more than ever to be excluded and any theoretical model implies the problem has been chosen' as was stressed by Prigogine and Stengers (1979).

Yet, so-called ecological studies, claim to escape that need for hypotheses. On the pretext that 'the paradigm determines the choice of questions we are asking ourselves, and thus the answers we want to get' (Rose *et al.*, 1974) for some people objectivity rules out any previous question, and thus reality appears in its virgin purity.

And then when overwhelmed by a jumble of observations, they rely on mathematical treatment to bring out what is important, and draw up correlations they formalize with regression lines which are both unrealistic and useless. Now Prigogine and Stengers (1979) have recalled that Hegel was highly creditable for 'showing that the possibilities for applying mathematical concepts to physical behaviours are restricted to the more ordinary ones'. It is through such aberrations that ecological research falls into disrepute.

Nicholson, who is considered as the father of population mathematical models, and to whom I wrote that this mathematization was impossible for the moment, whenever every physiological function can be modulated by the psychosomatic integration of environmental information, had answered me that for him, it was but a first step aiming at introducing experimental models in ecology, to do away with sentimental talking and writing. Such models were *reference models* for him, aiming at showing more complex phenomena. He wanted by no means to formalize descriptive models.

Therefore studying bruchid biology with an ecological viewpoint does not mean looking for mythical r and K parameters; it does not mean either comparing energetic yields to classify species 'in terms of their octane rating'.

To succeed in preventing bruchid attacks, there is no wonder recipe written on packings, or in entomological 'cookery textbooks' efficient all over the planet (Labeyrie, 1977).

One of the reasons accounting for the success of synthetic insecticides is that a lot of people thought they had found an all purpose powder capable of saving plants in any case; but agriculture cannot be supplied with such a wonder product.

There is not, there cannot be an all-purpose method to solve the problems of crop protection.

It is even impossible, *for a given insect*, to recommend *one* method of protection for *all* countries.

Our planet is heterogenous, ecological stresses vary from one area to another, the conditions of cultivation are never exactly the same, therefore the same protective method cannot be advocated in Ouagadougou or in Mexico, or even in Lima or Caracas.

For example the presence of corn as a companion crop has allowed a six or sevenfold reduction of the importance of *Acanthoscelides obtectus* attacks on bean pods in damp area, the 'Chalosse', in Southwest France; but 25 kilometers away in the north, in the sandy area of the Landes, the water deficit, which weakens corn, has removed the protective effect of the corn cover (Labeyrie, 1954).

Therefore, there cannot be a local protection without studying the ecological conditions beforehand. Indeed trying to economize in that field always proves to be expensive.

Any applied entomology textbook, which characterizes an insect without mentioning that the parameters are given as a guide – because they vary in time and in space –, is *dangerous*, and thus *responsible* for the mistakes and the inability of agriculture to control insect competition.

The living organisms, their populations are not interchangeable. While a car engine can be assembled exactly in the same way in Kano, Medellin or Detroit, crops and animals cannot be cultivated, reared and protected in the same way in various parts of the world.

BRUCHID AND LEGUME POLYMORPHISM SHOULD SYSTEMATICALLY BE TAKEN INTO ACCOUNT

Any individual, any population of living organisms shows, a certain plasticity allowing an adjustment to environmental fluctuations. Therefore, there exist no specific biological characteristics, nor a fortiori ecological characteristics (Labeyrie, 1969).

J.C. Headly (1972) rightly considers that 'agricultural pest control is still handled as though the pest were immobile and incapable of evolution'.

Unfortunately a large number of entomological works show that their authors have a typological point of view, according to Mayr's phrase (1974) and thus that they are by no means aware of the contingent value of results. Now in the biocenotic studies – and agricultural entomology can have but a biocenotic approach, as it studies the attacks of insects on plants –, polymorphism, at least at two levels (the plant and the insect) is generally overlooked. And the results yielded during a given season, in a precisely located field, get extrapolated in time and in space. Therefore, there are *no standard* 'ecological parameters' to be used in the laboratory for the study of bruchids. The conditions must simulate as much as possible the daily and seasonal thermoperiods and photoperiods of the areas populations were collected in.

Polymorphism is an insurance against the hazardous part of the environment; it allows populations to survive, by giving them a plasticity higher than the individual's. Bradshaw (1972) has remarkably indicated animal and plant capacities. His analysis is fundamental for any attempt to organize scientifically the protection of crops.

When confronted with a hazardous environmental disturbance, the animals can respond thanks to their behavioural possibilities, either by moving or by altering their behaviour. If this response is inadequate or impossible, their regulation systems induce homeostatis within their inner milieu. If these reactions are inefficient, the survival of population can be secured only by the survival of individuals whose plasticity was sufficient to absorb the ecological disturbance. Plants do not have an behavioural protection, and their capacities for physiological regulation are more limited. As a consequence their existence involves a higher hardiness and a morphological plasticity which provides a compensation.

Secundary -or allelochemical-substances (Fraenkel, 1959) of legumes, except perhaps those with a hypogean fructification (*Arachis hypogea*, *Kerstingiella geocarpa*, *Voandzeia subterranea*, . . .) play an essential part in their protection against bruchids (Applebaum, 1964).

Coleoptera Bruchidae whose larvae are specialist grain consumers (Utida, 1974) make up one of the very rare families which have colonized the ripe grains of legumes; the Bruchinae subfamily is even known on legumes only, according to Southgate (1979).

Owing to the specificity of legume allelochemical substances, the larvae of one Bruchid species often consume the grains of only one specie. For Janzen (1977), 73% bruchid species of arborescent Leguminosae of the deciduous tropical forest in Costa Rica attack the grains of one species, 15% two, 6% three . . . In High Volta, C. Varaigne-Labeyrie has found 63% species with one host and 25% with 2 hosts. Out of twenty attacked legume species, 16 were so by one bruchid species only. Johnson (1977) has found with *Desmanthus* pods in the U.S.A. the following distribution: one species with four bruchid species, 2 with three, five with 2 and five with one.

According to Bradschaw's model, the allelochemical content of legumes, and of grains in particular, varies with the ecological conditions. Bressani and L.G. Elias (1979) note that 'the environment affects the protein content and sulphur amino acids'.. Bliss (1979) considers that the heritability coefficient for grain proteins ranges between 0.25 and 0.60, which shows a very high phenotypic variance, due to ecological conditions.

Thus the chemical bond between bruchids and legumes must present a certain flexibility. The insect must be able to bear variations of the allelochemical content; its chemical adaptation cannot be absolutely rigid.

In addition, as the secundary substance composition is extremely polymorphic, nothing proves that all the grains of the same pod, all the pods of the same plant present both the same attractiveness and the same digestibility. For example, it would be interesting to determine the selective part played by gustation when *Acanthoscelides obtectus* feelers trail on the pods, in the sequence before boring the ventral suture.

Bradshaw (1972) consideres that protecting the structures which allow sexual reproduction is vital only in annual plants. In perenials, sexual reproduction is mainly an insurance against hazardous catastrophic accidents in the ecosystem and a means for colonizing free spaces.

This capacity for regenerating vegetative tissues, by forming new stalks, seems to be a general property of a large number of creeping or climbing legumes, which is utilised by farmers who mow clover or alfalfa several times on end. Genduso (1976) shows that this capacity for regenerating allows damage caused by *Bruchidius perparvulus* on *Trifolium alexandrium* to be avoided by harvesting the grains of the third cutting.

The indefinite habit and the capacity to form axiliary buds allow a large number of legumes to compensate for many destructions. Thus *Vigna unguiculata* could compensate for a 90% destruction of its flowers by neoformation (Simmonds, 1962).

For Evans (1974), the domestication of *Phaseolus vulgaris* has resulted in the selection of dwarf plants with a definite habit. In that case, the regenerating power has been removed. As a consequence, varieties coming from these selections can be seriously endangered as they have lost their regenerating power without having a set of chemical protection as sophisticated as those shown by annuals with a single fructification.

Fabre has remarkably described the situation of these plants altered by domestication: 'Such as they were made for man's requirements, unable to resist in the deadly struggle of living organisms, these plants would rapidly disappear, in spite of their vast number of seeds, if they were left, uncultivated, to their fate' (Fabre, 1879).

Therefore the results of experiments on legume attractiveness or sensitiveness cannot be generalized. The materials are so changeable and polymorphous that inferences should be made with the utmost care.

It is again because of the high polymorphism of this plant material that great efforts are being made to augment the nutritional value of legumes (by increasing the sulphur amino acid content) and their digestibility (by removing the inhibitors of trypsin and amylase, lectins, tannins, cyanogen glucosides).

The 'chemical standardizing', by removing specific substances, can lead to a reduced attraction of the more specialized insects, and thus perhaps makes it more difficult for them to discover the plants in the extremely complex chemical world of natural vegetation, or the agrosystems in which the chemicals released by other crops are prevailing. Now the movements determined by smelling are very important, not only univoltine bruchids such as *Bruchus pisorum* (Larson *et al.*, 1931) but also in polyvoltine species such as *Acanthoscelides obtectus* (Jarry, 1980). Conversely, this 'chemical standardizing' can favour the attacks of bruchids that were previously repelled; this is to be feared for lupin strains deprived of cyanogen glucosides. The production of chemically 'resistant' varieties can increase the selection, in specialist populations, of individuals favoured by an increased rate of allelochemical substances, as was shown by Stevens (1974).

Therefore any manipulation of secundary substances creates new ecological conditions by disturbing the olfactive and gustative world of biocenosis. It is likely to be as important as remodelling the landscape, even though our poor olfactory acuity makes it less perceptible to us.

Tegumentary protection can also be important. For example, *Callosobruchus maculatus* larvae are unable to bore the grain tegument of 19 legume species, and yet they can bore polystyrene 1 mm thick (Janzen, 1977). It is partly because thorough dessication reinforces the tegumentary barrier of grains that some peoples in South Columbia keep the pods above their hearths, even though it makes cooking more difficult later.

Now Pickersgill (1977) notes that: 'In *Phaseolus vulgaris*, both the parchment layer and the fibrous thickening of the sutures of the pods have been reduced, producing tardily dehiscent or indehiscent pods in the cultigens. The structure of the seed coat has also altered so that seeds of domesticated beans are more permeable, hence germinate more rapidly.'

It is obvious these selections make it easier for larvae to penetrate into the grains, but there may be other opposed effects. Jarry, for example, has observed that the dry pods of *Phaseolus vulgaris* varieties, whose texture had been considerably altered by selection were bored by *Acanthoscelides obtectus* females on ventral as well as on dorsal sutures. In a number of cases pod dehiscence even becomes abnormal, with a partial and localized opening of ventral suture. Now a local dehiscence favours penetration and oviposition of *Zabrotes subfasciatus* on the grains.

Other interventions make the protection of pulses much more complex.

Harlan reveals (1975) that 'the whole of soy bean industry is based on six introductions coming from the same Chinese area'. There is a consequent considerable reduction of its genic pool, and the disasters which swept through corn in the U.S.A., coffee and wheat in Brazil, tobacco in Cuba can occur again, and there is nothing to prevent the activity of bruchids from becoming the cause of similar disasters for soy.

Lastly, sensitiveness to insects is related to the plant's general condition. Now the varieties distributed by seed producers come from grains which have been selected and harvested in geographical areas which are quite different from the cultivation areas on the ecological point of view (Bradshaw, 1972). So cultivars, though they have been selected for their better resistance to insects and pathogens, often prove to be more sensitive in practice than original local varieties. Actually, *there is no capacity for resistance in the abstract.*

This principle is the more right as any selective factor, owing to its repetitiveness, determines evolution and allows adapted populations to develop. A few weevils are enough provided they are more resistant to colonize a disturbed system and initiate a new evolutionary process.

We can consider that the introduction of synthetic insecticides initiated a *very important radiation* in the evolutionary sense of the word. Using insecticides coming from allelochemical substances will not have to cause a new radiation; it will accelerate an evolutionary process which has been on for millions of years. That is the reason why we do not think that making chemically resistant varieties can bring a lasting solution.

THE BIOCENOTIC WORLD AND BRUCHID COMPLEXES WITH POLYTYPIC POPULATIONS

Coevolution is not a binary phenomenon in which two species at different trophic levels are involved alone.

The biocenotic universe is more complex; all the elements with a strong enough statistical relation can acquire an informative value, even if they do not play a direct part in the life of the respective species. For instance, when two plants have been associated in space in the course of their evolution, they are parts of the same information system and a herbivore can be expected to be able to locate its prey thanks to the characteristics of the associated plant, even if the latter is unfit for consumption (Labeyrie, 1978). This reasoning is valid at all levels and it is not systematically the consumed tissues or organs that must be attractive.

Various elements in the biocenosis can be operative in the meeting of sexes. Thus *Acanthoscelides obtectus* males and females get concentrated by the presence of *Phaseolus vulgaris* (Labeyrie, 1971) and it is from that time on that a male pheromone might operate (Hope *et al.*, 1967).

But all these phenomena do not have a timeless distribution. They all occur in a universe whose rhythms are imposed by latitude. Life on earth has been determined by the system's periodicity. Each population's evolution has taken place in systems with seasonal and daily periodicities characteristic of a given geographical situation, and of a given latitude in particular.

Thus any insect that is a specialist consumer of *Vicia faba* or *Pisum sativum* grains must have undergone an evolution, not only according to the respective allelochemical substances of these two legumes, but also according to their annual cycle. As no rhythm of vegetation can be as rigid as the only unalterable chronological indicator – photoperiod – the latter can set the insect in situation by causing alone or combined with thermoperiod and possibly with other associated periodicities, the annual activity to start again after imaginal diapause. But adjustment with the plant cycle requires a precision that cannot be achieved by photoperiod only. Thus coevolution may have introduced an element directly associated to the appearance of the favorable phenological stage. In the case of the pea-weevil, *Bruchus pisorum*, Pajni and Sood have shown that the formation of pollen is this indicator. It induces sexual maturity when it is consumed, and thus it allows females to have eggs at the right time to lay them on the young pods (Pajni *et al.*, 1975).

It is much more difficult to analyze the situation of the *Acanthoscelides obtetctus* complex. The utilization of certain ecological requirements as revealed by a precise physiological study, has enabled us to determine the area where the ancestors of bruchids infesting European beans were to be looked for. Let us recall the general outline of it. European populations of *A. obtectus* do not present an imaginal diapause. The males and to a certain extent the females cannot bear high temperatures; lastly, the presence of a canopy keeps the adults away. Under such conditions we have inferred that *A. obtectus* European populations came from an equatorial area since there is no seasonal rhythm; very likely at a high altitude, as they do not bear high temperatures; and of a savanna type, possibly with scattered trees, as the insects do not live under cover. As *A. obtectus* came to Europe three centuries after *Phaseolus vulgaris* was introduced, we decided to look for this insect in the Columbian piedmonts between 1200 and 2500 m.

Effectively, the conditions favourable to the permanent presence of *A. obtectus* or *P. vulgaris* crops are to be found over most of the Antioquia altiplano and the Nariño highlands as well as in the Cordillera valleys. In many regions, bean crops can be found at various stages of development and certain perennial Phaseolinae present indehiscent pods throughout the year.

In low areas, such as the Cauca and Magdalena valleys, high temperatures inhibit the presence of *A. obtectus*, but under such conditions, the grains are attacked by *Zabrotes subfasciatus.*

A. obtectus species is thus fractioned into allopatric populations favouring character differentiation. Huignard and Biemont (1978) have been able to note important physiological and ecological differences between populations which were collected at various altitudes. European populations present characteristics quite similar to those of populations in the Columbian piedmonts.

On the other hand, the populations picked from pods of *Phaseolus vulgaris* and *Phaseolus coccineus* crops in the fallow or abandoned lands of Tepoztlan, in the state of Morelos in Mexico are completely different (Biemont and Bonet, 1981).

These creeper *Phaseolus*, which are often to be found under a thicker canopy bloom and bear fruit in a short time early in the dry season; the reproductive organs of these plants do not appear progressively during their development. Evans (1979) records that *Phaseolus vulgaris* populations correspond either to plants with precise photoperiodic requirements, or to short day plants, or indifferent ones. They are the latter which were introduced in temperate areas.

A. obtectus attacks are very abundant on these Tepoztlan *Phaseolus* populations. But they are monovoltine insects which present a long imaginal diapause. On the other hand, in the same region, *Phaseolus vulgaris* crops are attacked by polyvoltine *A. obtectus*, apparently similar to those picked in the Columbian piedmonts or introduced into Europe. Thus in Tepoztlan, uncultivated *Phaseolus* have only one fructification period, like *Pisum sativum* or *Vicia faba*, and the monovoltine *Acanthoscelides* which attack them present an imaginal diapause like *Bruchus pisorum* and *B. rufinamus* which consume respectively the grains of the other two legumes.

The complex situation of the *Acanthoscelides obtectus* group shows it is necessary to study the insects in the dispersal areas of cultivated plants, and that the ecological observations, made in the highly disturbing conditions of cultivation which is often carried out far from the areas in which the plants were domesticated, cannot be sufficient.

The data on *Callosobruchus maculatus*' polymorphism (Utida, 1954) may presage the discovery of a similar complex situation.

Levins has stressed (1973) how important it is to analyze agrosystems to understand insect population dynamics; conversely, it is essential to examine the biocenotic relationships in wild plant populations and the crops in the areas of domestication of cultivated plants.

It is a matter of urgency that this should be undertaken for *Vigna unguiculata* in Ethiopia and *Glycine max* in the Far-East.

There is available information on the possibilities of attack of weevils on pods for only a dozen legume species out of the 89 man-consumed legume species. We have some rudimentary information on the ecology of only 4 or 5 harmful bruchid species.

This situation is extremely dangerous. The absence of attacks, on soy pods for instance, must not make us forget that the bean remained in a similar situation for several centuries in Europe, until *Acanthoscelides obtectus* was accidentally introduced a century ago. This is not an exceptional situation, for this insect seems to have been introduced in Chile in about 1937, *Bruchus rufinamus* seems to have joined broad bean in Japan in about 1920 (Kamito, 1933), *Tamarindus indica* and *Arachys hypogea* might have been attacked by *Caryedon serratus* in Mexico for ten years only (Bottimer, 1968).

POLYMORPHISM AND COLONIZATION OF STORED GRAINS

The colonization of stored bean grains involves a very simplified ovipositional behaviour (Labeyrie, 1964). By analyzing the stimulations which induce oviposition, Pouzat (1978) has shown the interferences of the tigmiotactism and chemotactism of the ovitube. The presence of adults of *Acanthoscelides* populations of the *obtectus* group, which stay after emergence in the pods of plants growing outside the crops in Tepoztlan, might imply that generations reproducing on grains in the pods have appeared; but these insects present an imaginal diapause. Actually it has not yet been possible to observe the reproduction of *A. obtectus* in situ after emergence in indehiscent pods.

Yet the idea cannot be rejected that the polymorphism of certain natural populations might allow females with lower stimulation thresholds to reproduce in indehiscent pods, or in lots of stored bean grains, which corresponds to a similar situation. The reproductive polymorphism of the populations which infest crops and stored grains in France is effectively conspicuous.

Some females, 10 to 20% of them, can release a few eggs without any stimulation induced by bean grains, when they are kept in empty Petri boxes after fecundation. Thus it is possible to select over a few generations 'astime' females (Labeyrie, 1961) whose nearly all descendants lay, without any trophic stimulation, large quantities of eggs. We have not there the emergence of a variety of domesticated weevils having lost their original biocenotic dependence. In effect, when placed again in the presence of grains, they show a higher ovarian production and an increased fecundity; these females are always capable of boring dry ripe bean pods.

Moreover, so far as the original trophic relationships are maintained and certain signals still contribute to the concentration of sexes, a sympatric speciation can hardly be considered (Labeyrie, 1978). But as the utilization of various hosts does not involve the same signals, the genic pools of fractions of the same population which colonizes various hosts may present various types of allelic frequences. The sympatric populations of *A. obtectus* in Tepoztlan seem to be much isolated from each other on the reproductive viewpoint, for polymorphism in their distinctive characters cannot be observed in their descendants; these populations are nevertheless slightly interfecund.

The bringing out of *Callosobruchus maculatus* polymorphism in nature might reveal different frequences of both types in the pods harvested in various ecological situations.

Likewise, the development of *Caryedon serratus* is not necessarily identical on its various Caesalpiniaceae hosts and on the *Arachys hypogea* grains it colonizes thanks to drying conditions which favour considerably infestation (Donahaye *et al.*, 1966). Some females of that species, like *A. obtectus* ones are very eclectic when choosing their laying site, as they can stick eggs on the surface of bottles (Appert, 1954). So we cannot reject the idea, a priori, that only a fraction of females would be attracted by *A. hypogea* pods and would oviposit on them.

The polymorphism of the *Bruchidius trifolii–B. alfierii*, complex which has been studied by A.K. Abou-Raya (1954) shows that the capacity for polyvoltinism is exclusively characteristic of the *B. alfierii* phase.

Therefore, we cannot think of achieving an efficient protection of legume crops without taking into account the remarkable polymorphism of bruchids. This polymorphism being poorly documented does not mean that it is limited to a few species, but that the biology of most bruchids has not been studied on populations collected in nature as a starting point.

Therefore weevil attacks could not be prevented by using only one technique: its repeated use would result in an oriented selection, favouring the phenotypes of weevils then capable of infesting crops (Labeyrie, 1977). For example, *A. obtectus* oviposits in pods which have reached a certain phenological stage; as long as dehydration of pods is not sufficient, the borings of the ventral suture by females are not followed by oviposition (Labeyrie, Maison, 1954b).

Thus, the sooner pods are harvested after their ripening, the smaller the attack is. Immediate shelling allows a further considerable improvement of protection by separating the eggs from the grains, before larvae can bore into them. The intensity of attacks is thus reduced fivefold (Labeyrie, 1957). But some females, about 10% in our observations, can oviposit in partly dehydrated pods. By contenting oneself with using this technique to protect crops, the risk might be run of favouring the females which accept to lay their eggs in incompletely ripe pods.

Likewise using the protection provided by a corn canopy in companion crops does not totally remove the attacks, since 38 weevils were obtained from 2960 grains picked under a corn canopy (Labeyrie, Maison, 1954a). Thus we can fear that a protection limited to that farming method may favour the females capable of reaching the pods in shady areas.

Wellington (1977) has stressed that whenever pest infestation is analyzed, it is indispensable to examine why some insects survive, because those are the ones which cause further failure. Thus when studying the methods preventing weevil attacks, it is indispensable to give the greatest importance to females which succeed in laying, and to the larvae that can develop.

Only the combined utilization of several protective methods can tend towards a permanent protection. In effect, by affecting various aspects of the insect's biology to disturb his biocenotic relationships with the attacked plant, various genic systems are being affected, with *independent* selection pressures whose effects can thus get added. It becomes highly improbable to be able to observe insects whose phenotype allows them to resist various independent selection pressures, which then makes the appearance of resistant population very unlikely to occur.

The close study of *Acanthoscelides obtectus*' biology has already allowed us to get very interesting results. The development of the biocenotic study of that insect will undoubtedly contribute to improve them. Similar studies of the other harmful weevils are absolutely necessary; they will allow simple methods to be devised which can be easily utilized by farmers, and provide an efficient protection.

But this implies the idea of a unique, world-wide protective method is given up. There are not, there cannot be such methods. It is by integrating various methods, resulting from a close knowledge of biology in determined biocenotic conditions, that an efficient protection can be insured.

The ecological study of bruchids must not only provide the means for protecting the legumes indispensable to mankind's protein food, but also an instructive example of intervention for agricultural entomology as a whole.

ACKNOWLEDGEMENTS

The observations in Latin America were made thanks to the help provided by the 'Centre National de la Recherche Scientifique' and in Mexico by the cooperation programme between the C.N.R.S. and the Mexican CONACYT.

REFERENCES

Abou-Raya, A.K., 1954. *Bruchidius alfierii* Pic a biologic race of *Bruchidius trifolii* MOTSCH. Bull. Soc. Fouad I entom., 38: 193–203.

Appert, J., 1954. La bruche des arachides. Bull. Agronom. STAT, 13: 181–90.

Applebaum, S.W., 1964. Physiological aspects of host specificity in the Bruchidae I General considerations of development comptability. J. Insect Phys., 10: 783–88.

Biemont, J.C. and Bonet, A., 1981. Observations préliminaires de populations de Bruches du groupe *Acanthoscelides obtectus* SAY vivant sur *Phaseolus vulgaris* et *Ph. coccineus* L. Sauvages ou subspontanés ou cultivés de la région de Tepoztlan (Etat de Morelos, Mexique). In: Ecology of Bruchids attacking legumes (pulses). Junk, The Hague, 23–42.

Bliss, F.A., 1979. Breeding legumes for nutritional quality. In: Adv. in Legume Science, pp. 179–85, (R.J. Summerfield and A.H. Bunting eds). Royal Botan. Garden, Kiew Pub., 667 pp.

Bottimer, L.J., 1968. Notes on Bruchidae of America North of Mexico with a list of world genera. The Canad. ent., 100 (10): 1009–49.

Bradshaw, A.D., 1972. Some of the evolutionary consequences of being a plant. Evolutionary Biol., 5: 25–47.

Bressiani, R. & Elias, L.G., 1979. Nutritional value of legume crops for human and animals. In: Adv. in Legume Science, pp. 135–51 (R.J. Summerfield & A.H. Bunting, eds.). Royal Botan. Garden, Kiew Pub, 667 pp.

Decelle, J., 1972. Les deux phases de *Callosobruchus maculatus*. Bull. Soc. Roy. Belge Entomol., 108: 27–28.

Donahaye, E., Navarro, S. & Calderon, M., 1966. Observations on the life cycle of *Careydon gonagra F.* on its natural hosts in Israel, *Acacia spirocarpa* and *A. tortilis.* Tropic. Science, 8 (2): 85–89.

Evans, A., 1974. Research on the evolution and genetic improvement of grain legumes. Depart. of Appl. Biol. Cambridge, 46: 5–11.

Evans, A., 1979. Structure, variation évolution and classification in *Phaseolus.* In: Adv. in legume science, pp. 337–47. (R.J. Summerfield and A.H. Bunting, eds.). Royal Bot. Garden Kiew Pub., 667 pp.

Fabre, H., 1879. Souvenirs entomologiques 8: 26–27, 47–48, ed. du Layet, 409 pp.

Fraenkel, G.S., 1959. The raison d'etre of secondary plant substances. Science 129: 1466–70.

Geier, P.W., 1966. Management of insect pests. Ann. Rev. Ent., 11: 471–90.

Genduso, P., 1976. Insectes nuisibles aux légumineuses en Sicile et observations sur l'hivernage des bruches univoltines. Boll. Ist. Ent. Agr. Oss. Fitop., Palermo, 10: 169–76.

Harlan, J.R., 1975. Our vanishing genetic resources. Science, 188 (4188), 618–21.

Headley, J.C., 1972. Economics of agricultural pest control. Ann. Rev. Ent., 17: 273–86.

Hope, J.A., Horler, D.F., Rowlands, D.G., 1967. A possible pheromone of the bruchid, *Acanthoscelides obtectus.* J. Stor. Prod. Res., 3: 387–88.

Huignard, J., Biemont, J.C., 1978. Comparison of 4 populations of *Acanthoscelides obtectus* from different colombian systems. Oecologia, 35: 307–18.
Janzen, D.H., 1977. How southern cowpea weevil larvae (Bruchidae: *Callosobruchus maculatus*) die on nonhost seeds. Ecology. 58: 921–27.
Janzen, D.H., 1977. The interaction of seed predators and seed chemistry. In: Comportement des insectes et milieu trophique, 415–28. V. Labeyrie, Ed., publ. C.N.R.S., 493 pp.
Jarry, M., 1980. Evolution et répartition des attaques par *Acanthoscelides obtectus* des gouses de *Phaseolus vulgaris* dans le Sud Ouest de la France. In: Ecology of Bruchids attacking legumes (pulses). Junk, The Hague, 133–43.
Johnson, C.D., 1977. Two new species of *Acanthoselides* from North-America and new host records from *Desmanthus* and *Hoffmanseggia*. The Pan Pac. Ent., 53: 60–73.
Kamito, A., 1933. On the broad bean weevil introduced in Japan. Proceed. of the Imper. Acad. Tokyo, 9 (3): 137–39.
Labeyrie, V., Maison, P., 1954a. De l'influence du microclimat sur la ponte de la bruche du haricot. C.R. Ac. Agric. France, 40: 733–35.
Labeyrie, V., Maison, P., 1954b. Sur les relations entre la ponte d'*Acanthoscelides obtectus* dans la nature et les stades phénologiques de *Phaseolus vulgaris*. C.R. Ac. Sc. Paris, 238: 1920–22.
Labeyrie, V., 1957. Influence des techniques de récolte des haricots secs sur l'intensité des attaques d'*Acanthoscelides obtectus*. C.R. Ac. Agric. France, 43: 138–40.
Labeyrie, V., 1961. Obtention d'une souche astime chez. *Acanthoscelides obtectus*. C.R. Soc. Biol. France, 155 (6): 1366–69.
Labeyrie, V., 1964. Importance de l'intégration des signaux fournis par l'hôte lors de la ponte des insectes. Année Psychol. 66 (1): 1–14.
Labeyrie, V., 1969. The variability of the physiological and ethological activities and the population growth of insects. Statistical ecology, 2: 313–36.
Labeyrie, V., 1971. Trophic relations and sex meetings in insect. *Acta phytopath.* Ac. Sc. Hungary, 6 (1–4): 229–34.
Labeyrie, V., 1977. For the definition of an ecological stategy in the protection of agrosystems. Experientia, 33: 404–10.
Labeyrie, V., 1978. Reproduction of insects and coevolution of insects and plants. Ent. exp. et appl., 24: 496–504.
Larson, A.O., Hinman, F.G., 1931. Some hibernation habits of the pea weevil in relation to its control. J. Econ. Eng. 24 (5): 965–68.
Levins, R., 1973. Fundamental and applied research in agriculture. Science, 181 (4112): 523–24.
Mayr, E., 1972. The nature of the Darwinian revolution. Science, 176 (4038): 981–89.
Mayr, E., 1974. Populations, espèces et évolution. Hermann pub. Paris, 496 pp.
Mitchell, R., 1975. The evolution of oviposition tactics in the bean weevil, *Callosobruchus maculatus*. Ecology, 56: 696–702.
Pajni, H.R., & Sood, S., 1975. Effect of pea pollen feeding on maturation and copulation in the beetle *Bruchus pisorum*. Indian J. exp. Biol., 13 (2): 202–203.
Pickersgill, B., 1977. Taxonomy and the origin and evolution of cultivated plants in the new world. Nature, 268 (5621): 591–95.
Poleman, T.T., 1975. World food: a perspective. Science, 188 (4188): 510–18.
Pouzat, J., 1978. Host plant chemosensory influence on oogenesis in the bean weevil *Acanthoscelides obtectus* Say (Coléoptère Bruchidae). Ent. exp. et Appl., 24: 601–608.
Prigogine, I., & Stengers, I., 1979. La nouvelle alliance, métamorphose de la science. Ed. Gallimard, 302 p.
Rose, S., & Rose, H., 1974. Le mythe de la neutralité scientifique. In: Responsabilité biologique, 221–30. Waston Fuller ed, Hermann pub., Paris, 260 pp.
Simmonds, N.W., 1962. Variability in crop plants, its use and conservation. Biol. Rev. 37: 422–65.
Singh, S.R., & Van Emden, H.F., 1979. Insect pests of grain legumes. Ann. Rev. Ent. 24: 255–78.;
Smartt, J., 1976. Tropical pulses. Longhman publ., 348 p.
Southgate, B.J., 1979. Biology of the Bruchidae. Ann. Rev. Entomol. 24: 449–73.

Stevens, N.E., 1974. How plant breeding programs complicate plant disease problems. Science, 184 (4135): 430–35.
Tropical legumes: resources for the future 1979. Nat. Ac. of Sc., Washington, 326 p.
Utida, S., 1954. Phase dimorphism observed in the laboratory population of the cowpea weevil, *Callosobruchus quadrimaculatus.* Japan J. appl. Zool., 18: 161–68.
Utida, S., 1974. Polymorphism in the adult of *Callosobruchus maculatus*, a possible process of evolution to stored product pest. Proceed. 1st Int. Workshop Conf. Stor. prod. ent., 689–91.
Wellington, W.G., 1977. Returning the insect to insect ecology: some consequences for pest management. Environnement entom., 6 (1): 1–8.
Whitehead, E.E., 1930. The pea weevil problem. J. of Econ. Ent., 23: 398.

Author's address:
I.B.E.A.S. (ERA CNRS 328)
Université F. Rabelais
Parc Grandmont
37200 Tours
France

UNIVOLTINE AND MULTIVOLTINE CYCLES

Their significance

B.J. SOUTHGATE
(Slough, Berks., U.K.)

INTRODUCTION

It is essential at the outset to define the two basic terms in relation to our own particular overall subject of bruchid ecology.

UNIVOLTINE (OR MONOVOLTINE)

The textbook definition of this is 'Having one generation per year or season'.

A species of Bruchus, living in the Mid Paleartic region and associated with a single crop, peas or beans, sown in spring and harvested in mid-summer ($3\frac{1}{2} - 4$ months later), provides a good example of a univoltine bruchid. The cycle being: eggs laid as soon as pods are initiated, the larvae developing with the growing seed, and adults emerging as the crop reaches maturity. Species following this life-cycle pattern are in sequence with a basic warm summer/cold winter climate and the adult hibernates during the winter. However in practice this simple cycle may be the exception rather than the rule.

MULTIVOLTINE (OR PLURIVOLTINE)

The alternative to one generation per year is a multiplicity of them. Only a few authors have considered this subject and they have assumed that more than one generation a year is a natural adaptation. I suggest that this is not necessarily so because if a perusal is made those species producing more than one generation a year it will be found that they are nearly all under the umbrella of 'economic species'.

They are all capable of infesting pods in the field, indeed, a certain proportion of them do so, the succeeding generations are completed in a man made environment 'the store'. All of these species appear at present to be totally connected with man's crops and infestation of wild plants is now a thing of the past. We may ask then, were it not for this artificial situation would these species have been univoltine or multivoltine. They have continued

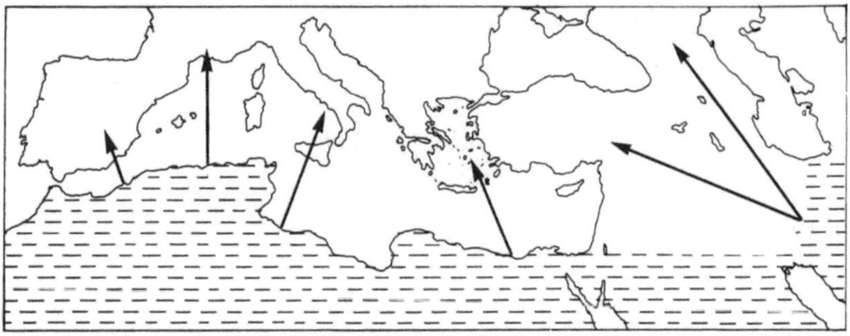

Fig. 1. Spread of Afro-Asian Callosobruchus spp., into southern Europe.

in this behavioural pattern even though they have spread into sub-tropical areas, helped by the agency of man (Fig. 1).

If we look back to the days of pre-agriculture we can assume that the wild ancestors of *Vigna* and *Cicer* species were plentiful enough to support a thriving population. The pods shattered on ripening and distributed seed to the ground. Assuming the seed was left where it fell and was not removed by being eaten by a large herbivore, the bruchid adult would emerge in due course. A female after mating will fly off in search of a freshly-ripened pod on which to oviposite. It is unlikely that all of the plants in a restricted area (e.g. within the flight range of a bruchid) will be in complete sequence. The odds then for the beetle reaching a freshly ripening pod would be fairly high and the cycle could begin again. But this situation would not last and eventually the pod source would cease. The beetle then, has two alternatives either it finds another host species or it develops an inbuilt adult diapause which will enable it to survive until the next season.

With regard to the two most troublesome economic species *Callosobruchus maculatus* and *C. chinensis* I think the alternative host was the system used, but this is pure speculation. Their multiplicity of generations is dependent on two basic conditions:

1. sufficient food availability, and
2. favourable climatic conditions.

What is of greater interest in the present context are the other members of the genus. Three further species are capable of breeding in store as well as the field. The remainder, some 15 or so described and 6 or 8 undescribed species as far as is known are 'field' species and incapable of continuous breeding. How many of these, given the opportunity could become storage pests. I believe the potential for multivoltism is inherent to a great number of bruchid species, as the evidence from the literature suggests. The present position in the 'economic' species I have summarised in Table 1. Material sent in for identification to Slough also supports this theory.

For example, during the study of the taxonomy of the Bruchidae associated with the *Acacia tortilis* complex. I have received some pods containing live material. This came from two main regions, 'The Negev' of Israel, and

Table 1. Numbers of species within genera of economic bruchids exhibiting univoltinism or multivoltinism.

	Univoltine	*Multivoltine*
Acanthoscelides	3	1
Bruchus	9	0
Bruchidius	3	1
Callosobruchus	6–8	5
Caryedon	1	1
Zabrotes	0	1

Africa (Kenya/Tanzania and the Western Central Sahara). The bruchid complex associated with it consisted of four *Bruchidius* and three *Caryedon* species. *Bruchidius* are the first to emerge and were at a much higher density and they were kept on the pods in the container in the laboratory. In the case of the material from Israel, pods of *A. tortilis* sub sp *raddiana* produced large numbers of adult *Bruchidius albosparsus* which gave rise to a second generation. The 'culture' then died out. The pods from Tanzania were *A. tortilis* sub sp *spirocarpa* and also produced *Bruchidius* species. This batch gave rise to a second and third generation of adults before dying out having been kept at 30 °C, 70% RH.

In the 'field' the bruchids oviposit at the later stages of pod development. The pod on reaching maturity normally drops to the ground. If the tree is growing in an arid environment then the pod merely drops into the tree canopy shadow, remaining there until conditions arise which are suitable to induce germination. The alternative is the removal of the pods or seeds from either the tree or the area beneath it by herbivores.

The late Dr Halevy of Hebrew University, Jerusalem, observed that a second generation of bruchids could develop in the field on fallen pods (Halevy, 1974). The absence of large herbivores seed predators gave rise to accumulation of seed and pods beneath the canopy thus enabling the beetles to continue for this second generation. Halevy did not arrive at any conclusion regarding the cessation of breeding by the beetles in the accumulated pods under the trees.

Under these circumstances then two main factors prevent the continuation of the species in this instance:

1. loss of moisture – therefore hardening of seed coat; and
2. absence of seeds

Further south in the African continent, particularly those with high animal populations, the seeds are not left below the trees for long periods as they are eaten very rapidly, therefore the possibility of further generations, developing are remote.

A further example is contained in the work of Mathwig (1971), (Fig. 2) in which he studied the populations of *Amblycerus robiniae* on *Gleditsia triacanthos* (Honey locust). Pods of the previous season's crop remain under the tree canopy from one season to the next, this enables adult females emerging from over-wintering sites to oviposit early in the season and before new season pods are available. The larvae from these eggs develops into adults within the month and there is a second generation following on immediately.

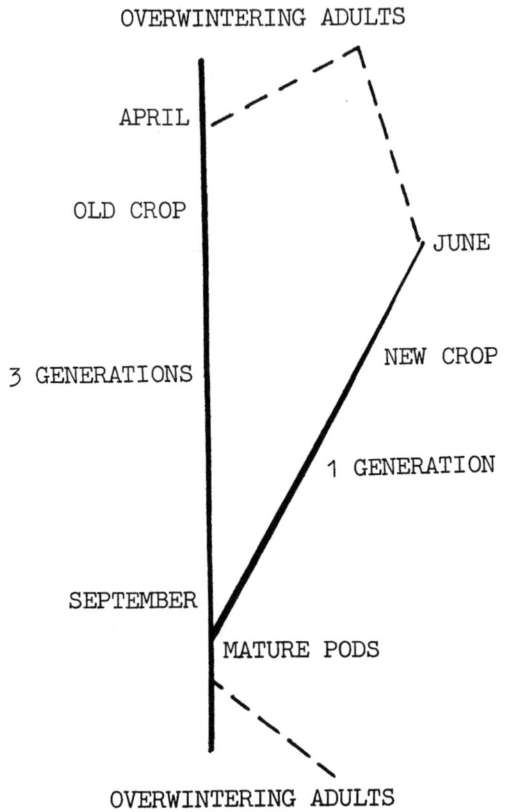

Fig. 2. Life cycle of Amblycerus robiniae on Gleditsia triacanthos (after Mathwig. 1971).

The third generation overwinters and so completes the cycle. A generation in new season immature pods develops concurrently with the 1st and 2nd generations on mature pods.

In these examples we have a progression from the univoltine to the multivoltine (via bi and trivoltinism). The question arises then – would the *Bruchidius* sp on the *Acacia tortilis* given suitable conditions develop into a truly multivoltine species? In other words what are the factors which govern these two arbitory states which the zoologist has defined? But more important still is, what are the advantages to the Bruchid to be one or the other? Before I attempt to answer that aspect I would like to summarize the factors which govern the continuation of the species under normal circumstances. These are:

1. availability of suitable hosts;
2. adaptability of the female to oviposition on different hosts;
3. larval tolerance of testa thickness and seed composition; and
4. absence of temperature regimes above and below the thresholds.

Given all of these conditions the species will reproduce normally.

These factors do not take into account the effect of parasitoides and predators neither does it attempt to bring in the manipulation of the host by man to vast unnatural mono-cultures often sequentially grown throughout the year.

One may ask whether it is advantageous to a species to exploit such vast food reserves as are provided by man. Undoubtedly survival of the species during unfavourable periods must be the main concern of the species.

Pest species in exploiting their unnatural food reserves face the disadvantages of large population build up, because:

a) Parasitoides can find hosts easily.

b) Food reserves are quickly used up and dispersion of remaining adults is necessary in order that the species may survive.

There are, then, a whole variety of factors involved in maintaining stability between these two states. We are far from understanding all the requirements of the individual species.

Having said that, it may seem to a non-bruchid specialist, that we have everything sorted out in the behaviour and ecological aspect of this group. Far from it. How many species are there, which at present are only capable of infesting hosts which are not really economic but could switch to other hosts which could be the economic crops of the future? Or to turn that question round the other way – how many hosts, at present already with bruchid associations, will be brought into cultivation in the future and thereby present a ready made problem. Let us take, what is perhaps an extreme example, but a nevertheless possible one – The Prosopis species of the New World and in particular those growing in the southern states of North America. This plant genera and in particular specific members of it, such as *P. velutina* and *P. juliflora* are possible candidates for resource management in certain semi-arid zones. They are already recorded as being infested with various bruchid species. Kingsolver *et al.* (1977) recorded 29 species belonging to 9 genera. Of these, at least one, is established as having 3 generations per year. If, as has been suggested by at least one worker, these were cultivated on orchard lines what will be the resultant seed crop? Figures quoted by Kingsolver *et al.* for normal 'wild' seed show that damage can be in the region of 55%. Given the ideal conditions of orchard type culture, which has been suggested, plus the development of improved strains these figures for damage could rise dramatically and thus make the use of some form of control essential right from the start.

What I am saying is this. I suggest that we should be looking at some of the factors which make up the basic biology of these other species. The question we must ask however is; How many of the fringe species, which at present, are not too troublesome providing they remain univoltine, can with just a little unconscious manipulation on our part, become multivoltine and pose an even more serious threat to the already seriously attacked legume crops? We shall, over the period of the next few days be hearing about plant relationships, allelochemicals, about oviposition response and a number of other topics regarding bruchid behaviour. But are we asking the right questions? Are we looking at the right aspects of bruchid/plant relationships? Our objectives are an understanding of the behaviour of the insect in order to achieve control by some means or other, chemicals, resistant strains, or agricultural practice.

Even at this stage we have little knowledge of the biology of even the common species in the 'field' let alone such fundamental facts as; do the species oviposit at night or during the day? – What are the threshold temperatures for flight? Is it essential that the adult feeds on pollen and from what source? Given answers to these questions we shall have hopefully a better understanding of bruchid behaviour in general and will be able to predict which species is going to become the most troublesome in the newer legume crops which are now coming into production, or how many of these 'low level' economic species which may become widespread in those crops already in existence and in which improved strains are being developed by the plant breeder.

REFERENCES

Halevy, G., 1974. Gazelles and seed beetles. Israel J. Bot. 23: 120–26.

Kingsolver, J.M., Johnson, C.D., Swier, S.R. & Teran, A., 1977. Prosopsis Fruits as a Resource for invertebrates, pp. 108–122. In: Mesquite its biology in two desert scrub Ecosystems (B.B. Simpson, ed.), US/IBP Synthesis Series 4. Dowden Hutchinson and Ross Inc. Stroudsburg, Pennsylvania, 250 pp.

Mathwig, J.E. 1971. Relationships between Bruchid beetles (*Amblycerus robiniae*) and Honey locust trees (*Gleditsia triacanthos*) PhD thesis. Kansas State Univ. Manhattan, 169 pp.

Author's address:
Ministry of Agriculture,
Fisheries and Food,
Slough Laboratory,
Slough, Berks.,
United Kingdom

THE BEAN WEEVIL POPULATIONS FROM THE *ACANTHOSCELIDES OBTECTUS* SAY. GROUP LIVING ON WILD OR SUBSPONTANEOUS *PHASEOLUS VULGARIS* L. AND *PHASEOLUS COCCINEUS* L. AND ON *PHASEOLUS VULGARIS* L. CULTIVATED IN THE TEPOZTLAN REGION STATE OF MORELOS – MEXICO

J.C. BIEMONT & A. BONET
(*Tours, France/Mexico, Mexico*)

INTRODUCTION

In Mexico, beans (*Phaseolus vulgaris* L. and *P. coccineus* L.) are widely cultivated. As in many other Latin American countries, their grains consumed when dry are the main protein source in human diet.

These crops are often cultivated by farmers who use traditional farming methods on small plots of land, in most cases in association with corn (Desir & Pinchinat, 1976). These grains, rarely treated chemically, are always infested by bean weevils resulting in a considerable destruction of harvested crops. On the whole in zones above altitudes of 1500 m only a very small number of *Zabrotes subfasciatus* is observed, whereas *Acanthoscelides obtectus* Say is abundant. Below an altitude of 1500 m, *A. obtectus* becomes rare and *Z. subfasciatus* abundant. In these low altitude zones, maximum temperature generally high, (> 35 °C) provoke the sterilization of *A. obtectus* males (Huignard & Biemont, 1974).

Today *P. vulgaris* and *P. coccineus* are cultivated on all continents. *A. obtectus* has succeeded in contaminating these crops in numerous countries all over the world. But ecological conditions and biocenotic contexts are very different from the original conditions. It is very difficult, therefore to study the adaptative value of behaviour and the cycle of the insects in habitats modified by man (Lack, 1965) by examining the biocenotic relationships between *Phaseolus* and *A. obtectus.*

Mexico is considered to be one of the probable zones where *P. vulgaris* and of *P. coccineus*, plants attacked in nature by *A. obtectus*, originated. *P. vulgaris* whose domestication in Mexico dates back more than 7000 years (Kaplan, 1965; Smith, 1968; Kaplan 1973) would have a North American origin, and more specifically speaking mexican for numerous authors (Kaplan, 1956; Miranda Colin, 1967; Miranda Colin, 1968, Gentry, 1968). But Kaplan *et al.*, (1973) indicate the presence of *P. vulgaris* cultivated in Peru.

Moreover Berglund-Brucher & Brucher (1976) consider *P. aborigeneus* (Bark), which originated in South America (Argentina) as the ancestor of cultivated *P. vulgaris.* In fact, numerous pre-columbian civilizations used beans and tried to domesticate them; the existence of these different zones of domestication makes the analysis of their origin difficult. According to Evans (1980) 'It is generally agreed that the cultivated species of *Phaseolus* originated in Latin America but whether from Middle America or South America is still debated'.

The status of these species is yet more complicated, since genetic isolation is not absolute. Indeed Miranda Colin (1966), Ibrahim & Coyne (1975), Evans (1980) indicate hybridization of *P. vulgaris* and of *P. coccineus.*

Wild or subspontaneous *P. coccineus* and *P. vulgaris* populations can be found in certain regions of Mexico. The biocenotic relationships between the *Phaseolus* and *A. obtectus* were examined in these regions where conditions are closer to the evolutionary context than are those of European bean crops.

For this study we chose a site rich in *P. vulgaris* and *P. coccineus* 2.5 km N.W. of the village of Tepoztlan (Servicio meteorologico Nacional 1976) in the Cuernavaca Region (State of Morelos-Mexico), 50 km south of Mexico City, 19° 00 north latitude and at an altitude of 1850–1900 m. (Fig. 1). The meteorological data supplied by the Cuernavaca station situated 20 km away and 300 m below indicate a rainy season, from about May–June to September–October. The temperatures recorded under shelter at Tepoztlan itself are shown in Fig. 2.*

1. STUDY OF POPULATION DEVELOPMENT OF *P. COCCINEUS* AND *P. VULGARIS* AND OF CULTIVATED *P. VULGARIS*

P. vulgaris and *P. coccineus* populations are liana forming and use trees, shrubs and branches as a support to climb. According to Raunkier's terminology (Cain, 1950), *P. coccineus* can be considered a cryptophyte and *P. vulgaris* a therophyte.

On the other hand, on small plots in Tepoztlan, we find *P. vulgaris* climbing stems of corn with which it is associated. These plots cultivated in a traditional manner, are immediately surrounded by abandoned areas where *P. coccineus* and *P. vulgaris* flourish. Each plot is harvested only once at the beginning of the dry season (November–December) and the harvest is conserved either in pod form, suspended or placed in sacks, or as stored seed.

In wild populations, the first sprouts appear in June–July and correspond with the germination of *P. coccineus* and *P. vulgaris*, or with the growth of *P. coccineus* from tuberose roots (Berglund-Brucher, Brucher, 1976; Baudet, 1977; Gentry, 1968). Thus, *P. coccineus* reforms two stems during the rainy season; Miranda Colin (1966) indicates the formation of a dense and widespread hairy like root. These plants then form an abundant canopy very rapidly, which in August 1979 covered the shrubs serving as a support. Their height may even exceed 4 to 5 m. Their stems are covered with numerous hairs used in the taxonomy of Phaseolinae (Gentry, 1968; Baudet & Marechal, 1976).

*In order to facilitate writing, when no further specifications are given, *P. vulgaris* and *P. coccineus* will be considered as being in their wild or subspontaneous forms, cultivated forms will be specified.

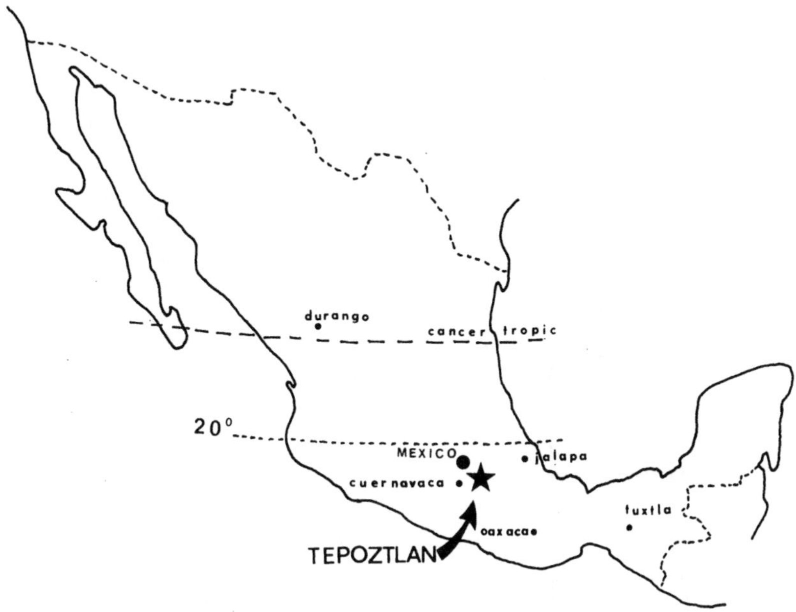

Fig. 1. Map of Mexico showing the geographic location of the area studied in Tepoztlan.

In 1979, the first flowers of *P. coccineus* were observed toward the end of August, whereas the beginning of flowering of *P. vulgaris* would be situated toward the middle of September. The first green pods of the two species are observed toward the middle of October. The pods reach maturity at the end of November, which corresponds with the beginning of the dry season (Fig. 2 and Fig. 5). The ripe pods of *P. vulgaris* are on the average 6 to 7 cm long and each contain an average of 8 viable grains. Seeds weight of wild *P. coccineus* and *P. vulgaris* are indicated in Fig. 3. These grains are 10 to 12 times smaller than the grains of cultivated local varieties. The ripe pods of *P. coccineus* are 5 to 6 cm long and an average of 6 viable grains.

Before total maturation, the *P. vulgaris* pods of these populations are characterized by the appearance of a slight opening along the ventral suture close to the peduncle. This opening exists neither in the cultivated varieties nor in *P. coccineus.* Finally the dehiscence of the pods provokes the projection of the grains of both species around the wild plants. By that time, these plants no longer possess green tissue, but bear only leaves and dry pods.

Some of these grains which have fallen to earth will germinate during the following rainy season. Although they fall to the ground in large numbers, they are difficult to find. This may lead us to suppose they were eaten by certain vertebrates.

In the laboratory, cultivated *P. vulgaris* grains germinate very rapidly (100% in less than 3 days) in saturated humidity at 20 °C (Fig. 4). However,

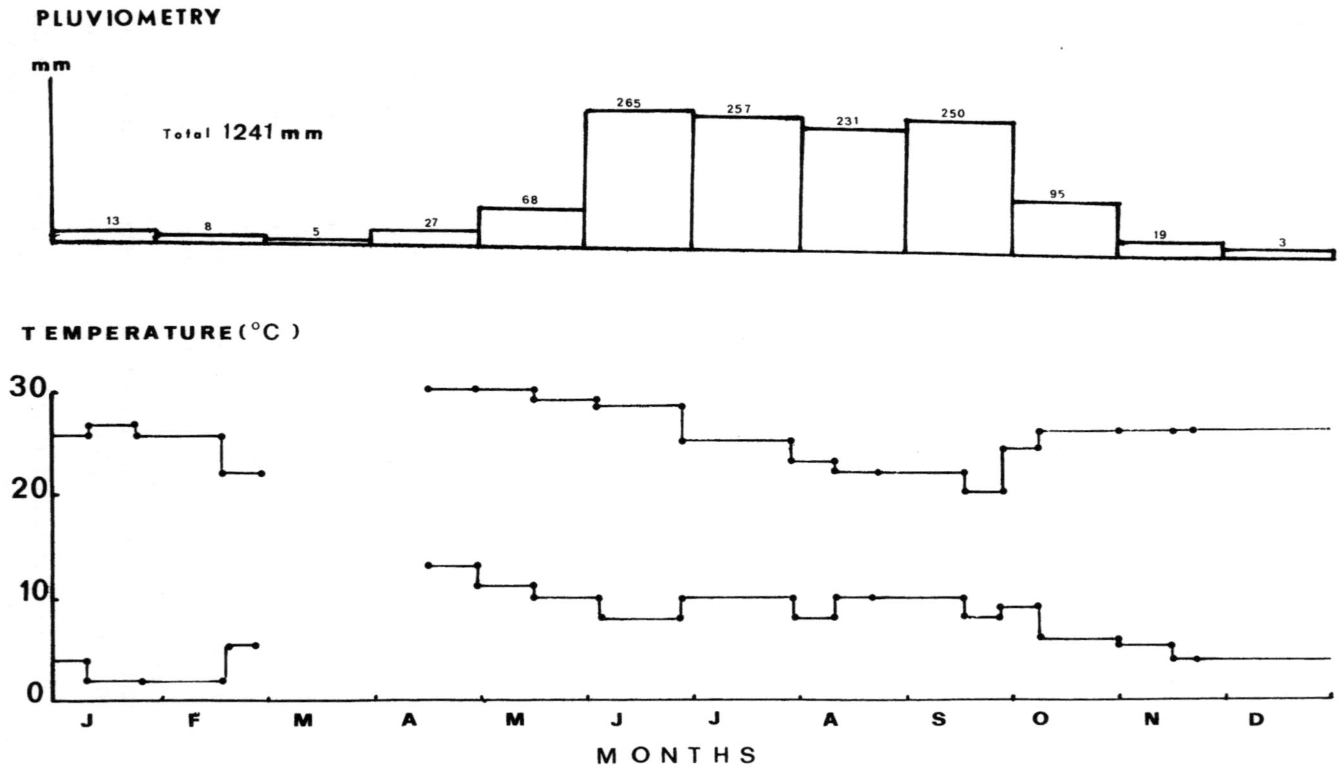

Fig. 2. Evolution of maximum and minimum temperatures under shelter, recorded at Tepoztlan in 1979, and the pluviometry at the Cuernavaca Station.

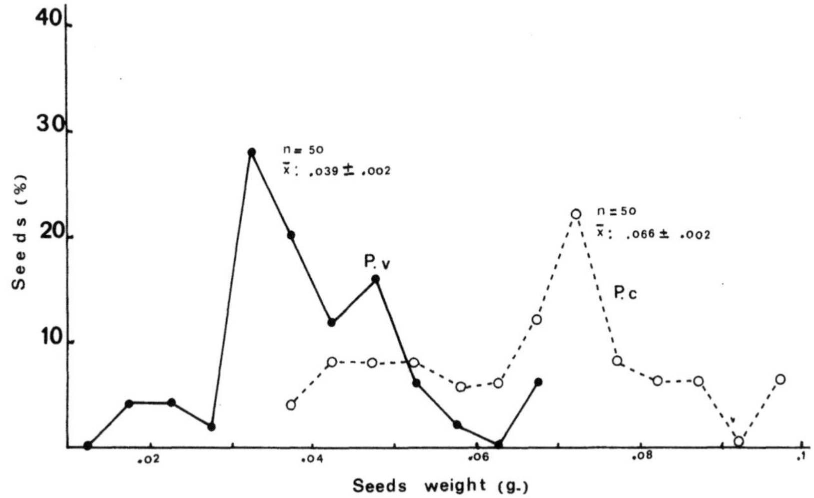

Fig. 3. Distribution of seed weight of wild *P. vulgaris* (P.v) and *P. coccineus* P.c) collected at Tepoztlan.

for seeds of wild populations placed in the same conditions, only 25% had germinated after 25 days.

For *P. coccineus* grains of these populations, there was only a 50% germination at the end of 25 days. However, after scarification of the tegument of *P. coccineus* and *P. vulgaris* grains, germination was as rapid as cultivated *P. vulgaris* (Fig. 4) which corresponds with Kaplan's observations (1956). We can therefore consider that the *P. coccineus* and *P. vulgaris* grains at Tepoztlan present a certain dormancy due to an inhibition of germination which would be caused by the grain's tegument.

2. CYCLE OF THE UNIVOLTIN BEAN WEEVIL ATTACKING WILD AND SUBSPONTANEOUS *P. COCCINEUS* AND *P. VULGARIS* (Fig. 5)

All bean weevil obtained from wild *P. coccineus* and *P. vulgaris* and from the cultivated *P. vulgaris* grains were identified by C.D. Johnson (personal communication) as belonging to the species *A. obtectus* Say.

During the entire period when ripe closed pods were absent, the adults were hidden in dry curled up leaves of *P. vulgaris* and *P. coccineus*, of other Phaseolinae or even of other plants. The leaves also constitute a shelter for about 40 other Bruchidae. The resting-behaviour in dry leaves was observed by Genduso (1978) in one univoltin type, *Bruchidius gilvus* in Sicily. Resting sites of the same type used by other univoltin species during imaginal diapause have been recorded. (De Luca, 1973; Prevett, 1967a, b). No adult was found in unrolled or partially rolled dry leaves, (*Quercus sp.* for example). Bean weevils only use dry leaves hanging from the plant. No adults were found in dry leaves on the ground. Increased humidity in the leaves on the

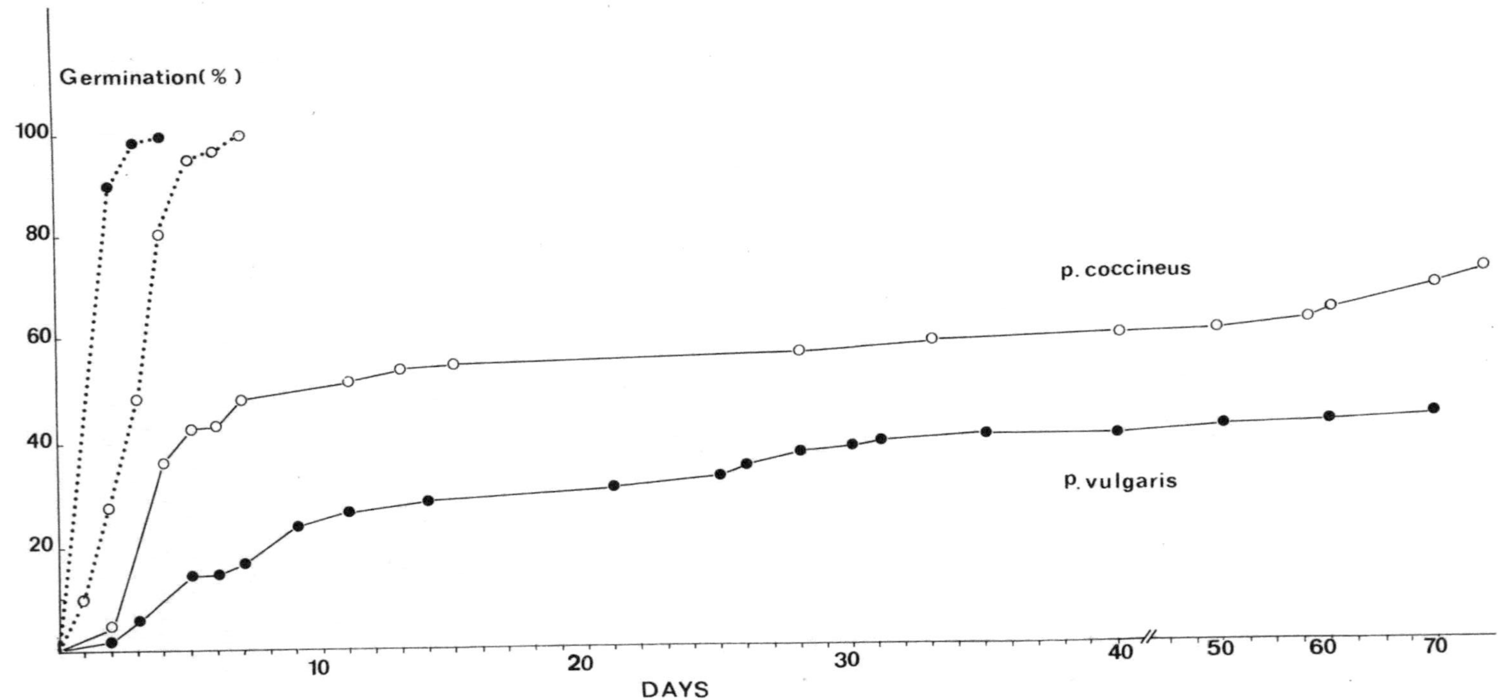

Fig. 4. Evolution of the percentage of grain germination in relation to time.
●: *Phaseolus vulgaris*; ○, *Phaseolus coccineus*; ———: grains intact; – – – –: scarified grains.
□ For each experiment, 10 repetitions, of ten grains each.

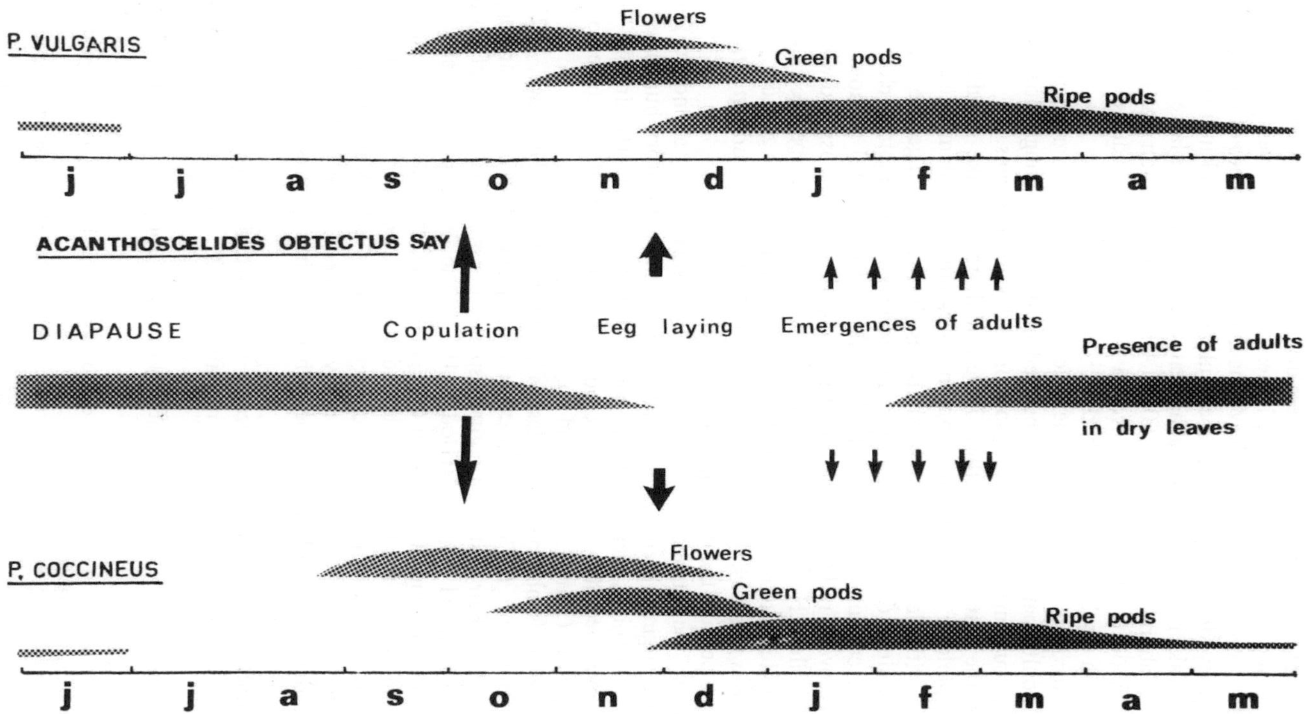

Fig. 5. Cycle of *Acanthoscelides obtectus* Say on *Phaseolus vulgaris* and *Phaseolus coccineus* in relation to the different phases of development of these plants during the year – Tepoztlan, Mexico.

ground does not completely explain why adults abandon them, because leaves attached to plants are often very damp during the rainy season. Bean weevils are not totally immobile in these dry leaves and they often move from one leaf to another. If we force the leaves to fall; the adult very rapidly fly away. Adults may occupy rolled leaves hanging from branches of cup plants. However, they do not use leaves accumulated and stratified in a suspended nylon net. Taken from their natural habitat, adults presented a large mortality rate in the laboratory (more than 50%). A similar observation was made by Pajni and Sood (1975) for the univoltin *Bruchus pisorum* adults who also find shelter in dry leaves or under bark.

Unlike of *Bruchus pisorum*, where females lay eggs in green pods, *A. obtectus* adults have never been observed in flowers, although flowers are quite numerous during the rainy season, or on developing beans.

During the rest period in these leaves, adult food intake should no be excluded. Mold or fungi may develop on leaves previously dampened and be consumed with decomposing vegetal tissues. In fact, the digestive tract of living maintained 6 to 7 months in *P. coccineus* pods, maintained closed by *Cuscuta* sp. contained a brownish black fluid. This survival without total immobility over a rather long period of time can therefore be assured without intake of pollen. However, it is impossible to state that a specific complementary food intake is not necessary for the maturation of the gonades.

In *Bruchus pisorum*, the specific ingestion of pollen from *Pisorum sativum*, for example permits the development of the gonades.

In August, the fat body of the adults of both sexes is very abundant. The females are virgin (absence of spermatophore in the copulatory bursa) until September–October; their telotropho-meroistic type ovarioles (Buning, 1972) are not developed, the very young oocytes having not as yet undergone previtellogenesis (Plate 1, Figs. A and B).

Unlike what is observed in males collected in Europe (Huignard, 1975; Cassie & Huignard, 1979) the annex tubulose glands of the males are not developed, but their testes do contain spermatodesms. The adults of both sexes are in imaginal diapause* (Hodek, 1973). The first copulations are not observed until the end of September or beginning of October, after regrouping in the same box in the laboratory the adults collected from the dry leaves. Moreover, one third of the females collected from the dry leaves at this time are fertilized but their ovarioles are rarely developed (Table 1). The imaginal diapause, therefore, ends in October when the first flowers of *P. vulgaris* appear in wild populations.

The adults collected from dry leaves in Tepoztlan in August 1979, and placed in the laboratory at a temperature of 25°–15° and 12 L–12 D,

*The use of the diapause concept is explained elsewhere by Hodek (1973–1980).

Plate 1. Fig. A: section of ovarioles collected from a female in imaginal diapause showing arrested ovarian development.

Fig. B: Histological section in a growing ovariole collected from a non diapausing female.

Fc = Folicular cells; G = Germarium; Yo = Young oocytes; Lo = Lateral oviduct; T = Trophocytes; V = Vitellus; G.v = Germinative vesicle.

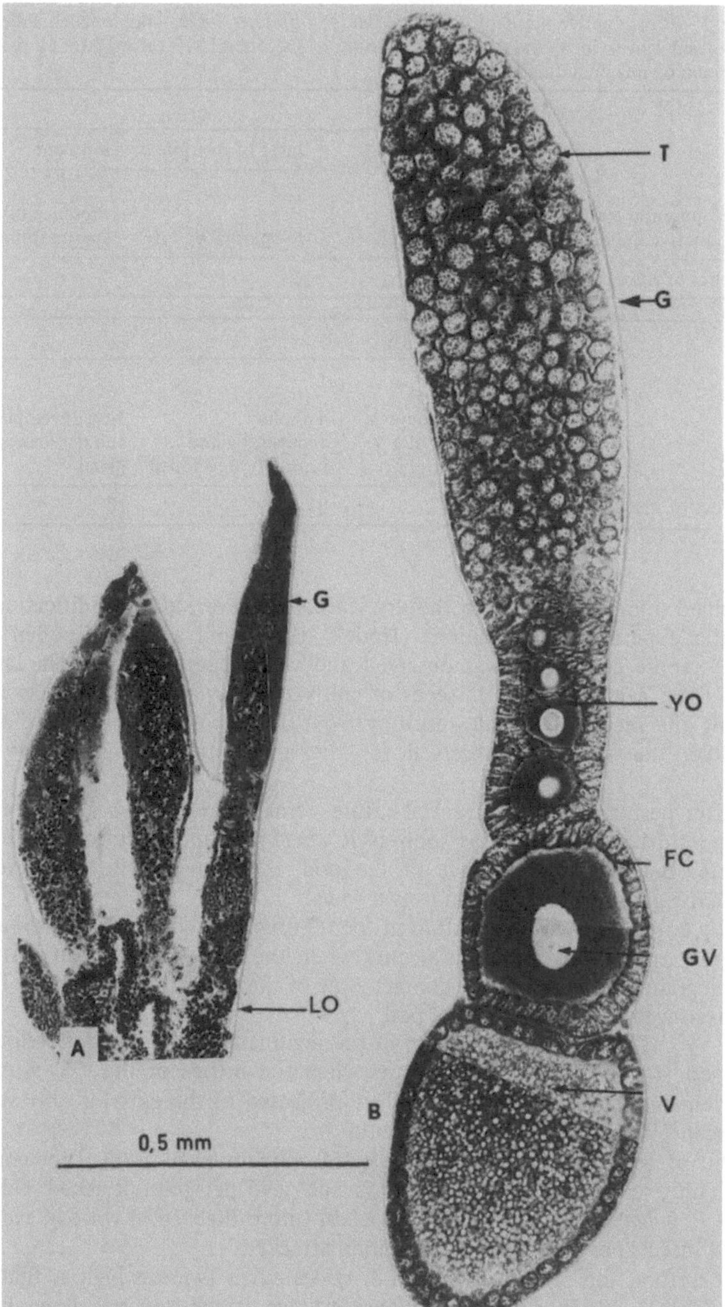

Plate 1.

Table 1. Reproductive state of *Acanthoscelides obtectus* males and females collected from dried leaves in Mexico, at the beginning of October 1979 (after 7 to 11 months maximum of imaginal diapause).

	Females, n = 28		
		Onset of ovariole development	
Developmental state of reproductive apparatus	Non growing No spermatophore	No spermatophore	Presence of a spermatophore
Number of insects	7	12	7

	Males, n = 18		
	Presence of mature oocytes Presence of a spermatophore	Tubular accessory gland normally developed	Non developed tubular accessory gland
Number of insects	2	8	10

remained immobile hidden in the dry leaves. However, after modification of specific experimental conditions (Hodek, 1980) and a diet of pollen and honey, some copulation was observed and viable eggs obtained. The larvae from these eggs developed in seeds of cultivated *P. vulgaris*. Two weeks after leaving the seeds, no female resulting from these crosses showed developed ovariole; the same was observed for the annex tubulose glands of the males.

After perforation of the ventral suture, females lay their eggs in clump as on the cultivated varieties as soon as *P. coccineus* and *P. vulgaris* pods are mature. Before the dehiscence of the pod, the bean weevils that develop inside the grains can leave them in two ways:

– by cutting out an operculum in the tegument of the grain as well as in the pericarp of the pod. This occurs when the orifice of the exit is made on the grain at the level of the contact zone of the pericarp and the pod. Thus the bean weevil is freed from the pod.

– by cutting out an operculum in the tegument of the grain as well as in the pericarp of the pod. This occurs when the orifice of the exit is made and remains inside the pod and will be projected to the exterior only when dehiscence takes place.

Out of 1631 grains of *P. coccineus*, 459 were attacked (28% of the grains) with, on average, .55 insects per grain and 1.95 per grain attacked. Out of 2569 *P. vulgaris* grains, 396 were attacked (more than 16%) with an average of .24 insects per grain and 1.45 per grain attacked.

Therefore, the rate of attack for *P. coccineus* is twice as high as that for *P. vulgaris*, with in addition, a higher average attack rate per grain. Leroi and Jarry (in preparation) were able to ascertain that possibilities of egg-laying and development of *A. obtectus* (European strain) are better on *P. coccineus* than on *P. vulgaris*.

Certain smaller and more irregular perforations correspond to exit holes of hymenopters parasites. They belong to five different families (Cheloninae, Eulophidai, Eupelmidae, Eurytomidae, Pteromalidae) currently being identified by Steffan (1980).

3. CYCLE ON *P. VULGARIS* GRAINS STORED IN TEPOZTLAN

In all the groups of *P. vulgaris* grains cultivated by farmers, the *A. obtectus* adults observed in August (that is during the diapause period of adult from the wild pods) were sexually active. The females were often fertilized, with ripe oocytes retained in the lateral oviducts, the accessory glands of the males were developed.

Thus bean weevils whose modes of reproduction differ, live in sympatry. In fact, in the first case, the cycle of the males and females with imaginal diapause, corresponds, in the wild Phaseolus populations, to the period where pods are absent (at the end of the dry season and the rainy season); in the second case, the reproductive cycle is linked with the possibilities of continued development all year long, on stored grains or on the crops during the maturation period of the pods. This shows the complexity of analysis of *A. obtectus* development in Tepoztlan, where both uni and pluri voltin populations cohabited.

4. COMPARISON OF UNIVOLTIN BEAN WEEVILS FROM WILD POPULATIONS IN TEPOZTLAN WITH THOSE OF A STRAIN COLLECTED IN FRANCE

In all the females from a strain collected in France bred in the laboratory, the lateral oviducts which permit the retention of mature (Biemont, 1975; Biemont, 1979a) oocytes have a thin epithelium and the cells located toward the light at the base of this oviduct are lined with chitinous patches (Plate 2, Fig. A and B). In females collected from dry leaves we observe directed toward the light in the lateral oviduct, long filaments which seem to be cellular prolongations of the epithilium surrounded by a thin chitinized intima (Plate 2, Fig. C and D). It is interesting to note that the insects coming from stored grains from the State of Durango, present yet a different structure at this level (Plate 2, Fig. E and F).

The structure of the rosette surrounding the micropyle of the *A. obtectus* eggs collected in Europe (Biemont & Huignard, 1979; Biemont *et al.*, in preparation) is composed of between 8 and 15 petalloidal forms in contrast with 3 for the univoltin female eggs from Tepoztlan (Plate 3, Fig. A and B). Immunodiffusions carried out between an antiserum prepared against egg vitellin protein (Biemont, 1979b) and hemolymph of *A. obtectus* coming from beans stored in Mexico revealed the existence of a precipitation arc analogous to that which was obtained with hemolymph from French female strains (Plate 3, Fig. C).

The vitellin protein in the females originating from Mexico, is therefore immunologically identical to that of the French females. A similar result was obtained with *A. obtectus* females from Columbia (Huignard & Biemont, 1979). However, this technique did not enable us to show the presence of this protein in diapausing females from the dry leaves from Tepoztlan. This protein does appear after the imaginal diapause has been completed.

5. GENETIC EXCHANGE POSSIBILITIES IN UNIVOLTIN *A. OBTECTUS*

All authors have indicated that males and females from stored grains copulate and give rise to diapause free descendents.

When, after termination of their imaginal diapause, univoltin strain males from Tepoztlan are crossed with females of other strains, different situations appear.

When univoltin females from Tepoztlan are fertilized, after termination of their diapause, we obtain F_1 which are all in imaginal diapause.

When males of the univoltin strains from Tepoztlan are copulated with females of a European strain, the reproductive systems of the F_1 males and females are not in the same stages of development (Table 2).

Table 2. Results of European strain female crossing with males originating from Tepoztlan (after termination of diapause; Hodek, this volume).

Females		Males	
Ovariole development		Tubular accessory gland development	
+	–	+	–
27	14	36	10

Ten out of 46 males did not possess a developed annex tubulose gland. Only 27 out of 41 females were able to copulate and lay viable eggs. Fourteen neither copulated nor laid eggs. One month after emergence, their ovarioles remained undeveloped. They no longer had a discernable fat body and their lateral oviducts were swollen by a secretion from cells of the walls.

No F_2 from these mature females presented diapause. Crosses between these F_2 produced F_3.

CONCLUSIONS

Plant susceptible of being consumed by *A. obtectus* in the Tepoztlan area of Mexico, are highly heterogeneous. Germination in wild or subspontaneous *P. vulgaris* is obtained for certain grains only after a long exposure to a humid atmosphere. However, the germination of *P. coccineus* is more rapid which could be unfavorable for the plant as germination obtained after an accidental light rainfall would not permit development should a dry spell follow. The presence of a tuberous root allowing it to regenerate new sprouts after the dry season, relativizes the possible loss of grains due to a very early germination or to attacks of *A. obtectus* or other insects. Thus, sexual reproduction could be of less importance in comparison with the possibilities of regeneration from the mother plant. The situation is totally different for the cultivated or wild populations of *P. vulgaris* for which sexual reproduction is the only possibility. The mother plant cannot survive during the dry season. In cultivated *P. coccineus* and *P. vulgaris*, selection had favoured rapid germination.

Therefore, on one hand, we have *P. vulgaris* and *P. coccineus* forming populations with fruitifications at the beginning of the dry season, and grains presenting a certain dormancy, and on the other hand, crops of domestic varieties of *P. vulgaris*, with maturation at approximately the same period and conservation of their pods and grains by man.

The host plant of *A. obtectus* therefore shows two different modes of reproduction: an annual cycle of wild populations with ripe pods and grains available for a short period of time, during the dry season; and possibilities for germination all year long for stored grains of the cultivated plants which are always available.

Two types of bean weevils presenting different reproductive particularities develop on these plants whose status are quite different. Bean weevils from wild populations of *P. coccineus* and *P. vulgaris* are univoltin with imaginal diapause when ripe pods are absent; those developing on stored grains are polyvoltin without imaginal diapause. The absence of ripe pods and of grains in the wild populations leads to an adaptation by imaginal diapause characterized by:

– absence of ovariole development, presence of an abundant fat body, mobility, and possibility of food-intake by the female;

– absence of the development of tubulose accessory glands, presence of spermatozoids in the testes, abundant fate body, mobility and possibility of food-intake by the male.

These are the criteria of an imaginal diapause (Hodek, 1973; Tauber & Tauber, 1976; Boiteau *et al.*, 1979; Masaki, 1980) which permits an adaptation to periodic ecological conditions.

In this particular case, it is a question of synchronising the reproductive capacities of the insect and the presence, in wild populations, of the phenologic stage of the host plant necessary for egg-laying and the development of the larvae.

The synchronization between one insect stage and the presence of food necessary for development of that stage is one of the adaptative functions of diapause. During the diapause, the sexual activity of the adults is momentarily suspended, however they are not lethargic and can show flight activity. Such activity during diapause may be energetically unfavorable, but it would, in fact, permit the insects to avoid their destruction as well as the destruction of their shelters which are not permanent (dropping to the ground of the dry leaves serving as shelter).

This constitutes an adaptation to a variable environment. An adaptative

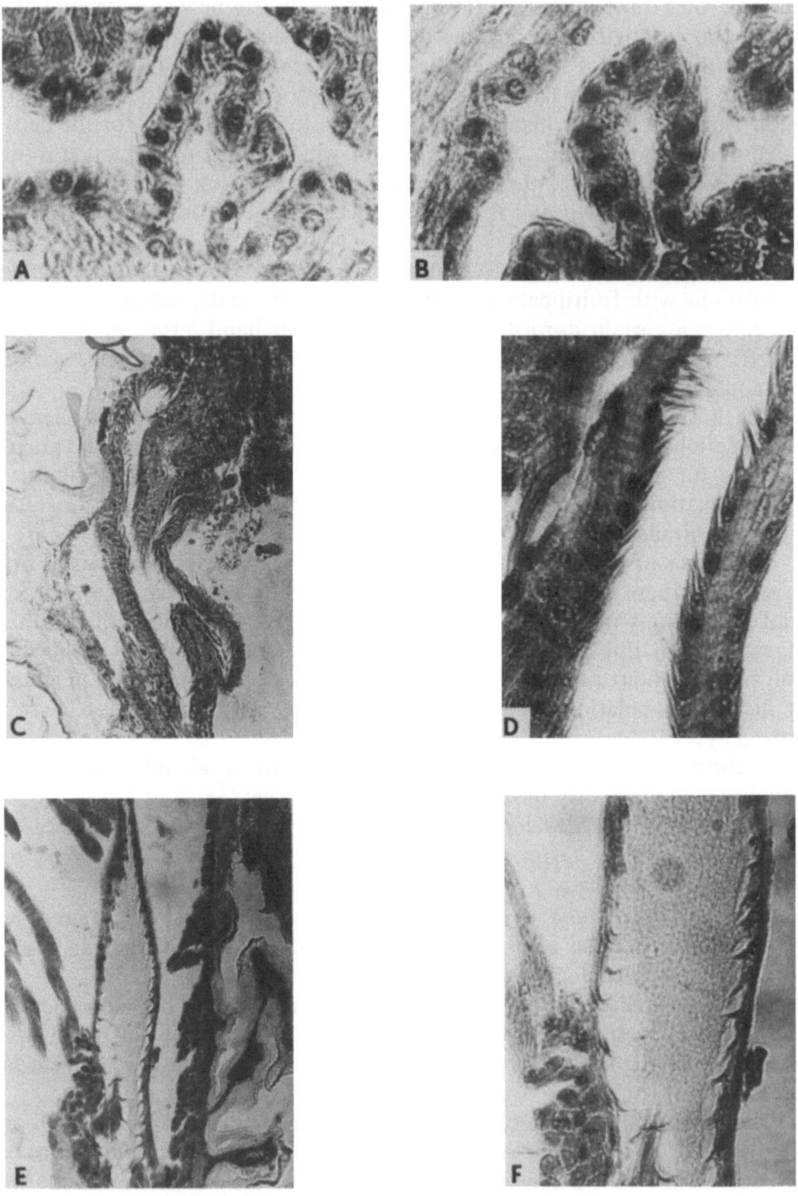

Plate 2. Intima chitinuous structure located at the base of the lateral oviducts of *Acanthoscelides obtectus.*

Figs. A, B: *Lateral oviduct base section* of a European strain female. The epithelium seems to be covered with chitinous patches.

Figs. C, D: Base of the lateral oviduct of diapausing females from Tepoztlan.

Figs. E, F: Base of the lateral oviduct of females from stocks in the Durango region of Mexico.

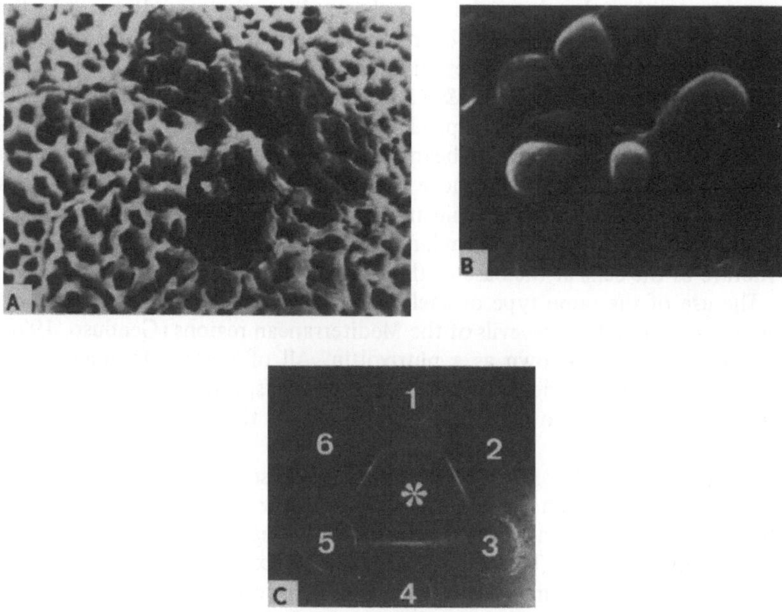

Plate 3. Fig. A: Micropylar region of a mature oocyte taken from lateral oviduct of a female collected in wild population at Tepoztlan.

Fig. B: Micropylar region of a mature oocyte of a European strain female.

Fig. C: Immunodiffusion (Ouchterlony technique). *Anti-vitellin protein serum; 1) Extract of an adult European strain female; 2) Extract of an adult female collected in stored beans; 3) Extract of a diapausing female: vitellin protein is not detected; 4) Extract of mature oocytes of a European strain female; 5) Extract of male (vitellin protein is not present); and 6) Hemolymphs of an adult European strain female.

mechanism of the same order noted by Morton (1971) in his observations of tropical birds.

Each one of these two types, therefore, has adapted to the reproductive particularities of the wild population or of the cultivated plant modelled by man who makes the grain permanently available to the insect in storage. One may wonder if the insects aptitude to colonize stocks is not linked with the particularly complex polymorphism of its reproduction.

In polyvoltin bean weevils, favorable conditions for development are always present. Bean weevils collected on pods and isolated grains stored by the small farmers, were sexually active (female having copulated and containing mature oocytes; developed accessory glands in the males). Continuous breeding can be obtained from these adults.

Thus two populations, one univoltin on wild populations, the other polyvoltin on cultivated plants are sympatric. This makes the study of *A. obtectus* on *P. coccineus* and *P. vulgaris* in Tepoztlan, more difficult.

There may be a certain reproductive isolation between the populations at the time when bean weevils on wild populations are in diapause. But this isolation is not permanent because the adults of these populations may reproduce with polyvoltin adults at the end of diapause. If reproduction is possible between polyvoltin bean weevils and those of wild populations during their period of sexual activity, would this be similar to a differential choice of partners as (Watanabe & Kawanishi, 1979) indicates for Drosophila.

We would then find at the reproductive level, a certain polymorphism that we have not observed with the bean weevils developing on stored grains. The difficulty is increased all the more by the impossibility of discerning these two types using classical taxonomic methods. However, a certain polymorphism may be observed at the micropylary rosaceous level of the egg, and in the structure of the cells at the base of the lateral oviduct.

The use of the same type of shelter that was observed in Tepoztlan exists with the univoltin bean weevils of the Mediterranean regions (Genduso, 1978). Yet, *A. obtectus* is known as a plurivoltin. All of these criteria and observations lead us to wonder if this really is the same species. Several hypotheses may be put forward to explain the existence of the sympatry observed in Tepoztlan.

In Mexico, the *A. obtectus* exhibiting diapause, develops on *P. vulgaris* and *P. coccineus* which present a certain dormancy. However, in Columbia, where the host plant is available to the insect all year round, *A. obtectus* does not exhibit diapause. The different types of exchange would then have brought polyvoltin bean weevils who would have colonized the cultivated plants. The coexistence of these two types of reproduction would then be maintained. The existence of diapause, which implies a more complex hormonal system than that of the continuous development type bean weevils, could be a more recent evolutionary phenomenon than the absence of regulation. Under these conditions we must accept that certain polyvoltin bean weevils would derive from univoltin bean weevil originating in equatorial regions where they can develop on beans present all year long. We may also suppose that during the period of sexual activity, univoltin bean weevils could have developed on the cultivated plant and then colonized the grains stored in Mexico. By exchanges, these bean weevils would be widespread wherever the grain available all year round would permit its development.

Because of their richness, *P. coccineus* and *P. vulgaris* grains constitute the essential protein contribution of vegetal origin for numerous human populations, in Mexico, in particular, where numerous different varieties of *P. vulgaris* are grown by small farmers for their own consumption. Small subsistance agriculture is very widespread among rural populations.

As with many legumes consumed or consumable by man (National Academy of Science, 1979) *P. coccineus* has received but little attention.

Moreover, *P. coccineus* and *P. vulgaris* eating insects were officially considered only as insects of stored grains. The conditions of crop contamination had not yet been studied.

All of our observations show the complexity of the insect, plant-host relationships particularly when a polymorphism exists both at the insect and plant level.

Therefore, one must be particularly careful before drawing conclusions and making generalizations from individual observations. It is impossible to extrapolate from observations made in one specific country in order to define methods of control in another. Protection of a plant against an insect involves interrupting the relationship between the insect and the host plant, it is impossible to achieve this if we do not know the nature of these relationships perfectly well, unless we continue using blind methods which have proved their noxiousness.

ACKNOWLEDGEMENTS

This work was done in agreement with the C.N.R.S.–CONACYT. Institut de Biocenotique Experimentale des Agrosystèmes de Tours (France) and the Instituto de Ecologia de Mexico (Mexico), Dirección General de Investigación Científica y Superación Academic de la Subsecretaria de Educación Superior et Investigación Científica, de la Secretari de Educación Pública.

We would like to thank C.D. Johnson for determining the Bruchidae, J.R. Steffan for determining hymenoptera parasites of bean weevils, A. Delgado and R. Marechal for determining Phaseolinae, and I. Hodek, M. Jarry, B. Leroi and J. Pouzat who also collaborated with this work both in the fields and in the laboratory.

REFERENCES

Baudet, J.C. & Marechal, R., 1976. Signification taxonomique de la présence de poils uncinulés chez certains genres de Phaseoleae et d'Hedysareae (Papilionaceae). Bull. Jard. Bot. Nat. Bel. 46: 419–426.

Baudet, J.C., 1977. Origine et classification des espèces cultivées du genre *Phaseolus* Bul. Soc. Roy. Bot. Belg. 110: 65–76.

Berglund-Brucher, O. & Brucher, H., 1976. The south american wild bean (*Phaseolus aborigeneus* Burk.) as ancestor of the common bean. Econ. Bot. 30: 257–272.

Biemont, J.C., 1975. Influence des modifications de courte durée de la température d'élevage sur la capacité reproductrice de la bruche du haricot *Acanthoscelides obtectus Say* (coleoptera: Bruchidae). In: Sterility principle for insect control. IAEA 6: 279–288.

Biemont, J.C., 1979a. Influence de la plante-hôte et de la copulation sur la levée de l'inhibition du développement ovarien liée à la rétention des ovocytes chez *Acanthoscelides obtectus* (Col. Bruchidae). Ann. Soc. Ent. Fr. 6: 999–1006.

Biemont, J.C., 1979b. Vitellogenesis in *Acanthoscelides obtectus* (Coleoptera: Bruchidae): I Oocyte development and vitellogenin in a european strain. Int. J. Invert. Reprod. I: 221–232.

Biemont, J.C. & Huignard, J., 1979. Diversité des structures micropylaires des ovocytes d'*Acanthoscelides obtectus* Say (Coleoptère Bruchidae). Atlas de microscopie à balayage (123–124) Librairie Arnette.

Biemont, J.C., Chauvin, G. & Hamon, C., L'oeuf d'*Acanthoscelides obtectus* Say (Coleoptère Bruchidae). Ultrastructure et résistance à la dessication. (in preparation).

Boiteau, G., Bradley, J.R. & Van Duyn, J.N., 1979. Bean leaf beetle: some seasonal anatomical change and dormancy. Ann. Ent. Soc. Am. 72: 303–307.

Buning, J., 1972. Untersuchungen am Ovar von Bruchidius obtectus Say (Coleoptera polyphager) zur Klärung des oocytenwachstums in der Prävitellogenese. Z. Zell forsch. 128: 241–282.

Cain, S.A., 1950. Life form and phytoclimate. The Bot. Rev. 16: 1–32.

Cassie, P. & Huignard, J., 1979. Etude ultrastructurale des glandes annexes de l'appareil génital mâle chez *Acanthoscelides obtectus* Say (coleoptera: Bruchidae). Int. J. Insect Morph. Embryol. 8: 183–201.
Cetenal 1976. Carta topografica Secretaria de la presidencia, Mexico.
De Luca, Y., 1973. Comportement d'inactivité imaginale de Bruchides. (col.) Annls. Soc. Mort. Hist. Nat. 113: 79–83.
Desir, S. & Pinchinat, A.M., 1976. Producion agronomic y economica de maiz y frijol comum as ociados segun tipo y poblacion de plantas. Turrialba 26: 237–240.
Evans, A.M., 1976. Beans Phaseolus spp. (Leguminosae: Papilionaceae). In: Evolution of crop plants (168–172). Simmonds, N.W. (ed.), 333 pp.
Evans, A.M., 1980. Structure, variation, evolution and classification in *Phaseolus.* In: Advances in legume Science (R. Summerfield & A.H. Bunting, eds.).
Genduso, P., 1978. Insectes nuisibles aux légumineuses en Sicile et observations sur l'hivernage des bruches univoltines. Boll. Int. Ent. Agr. Oss. Fitopat. Palermo 10: 169–176.
Gentry, H.S., 1968. Origin of the common bean *Phaseolus vulgaris.* Econ. Bot. 23: 55–59.
Hodek, I., 1973. Biology of Coccinellidae. Ed. Academia Prague, 260 pp.
Hodek, I., Bonet, A. & Hodkova, M., 1980. Some ecological factors affecting diapause in adults of *Acanthoscelides obtectus* from Mexican mountains. Ecology of Bruchids attacking legumes (pulses). Junk, The Hague.
Huignard, J. & Biemont, J.C., 1974. Influence d'une augmentation de la température sur la capacité reproductrice de mâles chez la bruche du haricot *Acanthoscelides obtectus* Ann. Zool. Ecol. Anim. 6: 561–574.
Huignard, J. 1975. Anatomie et histologie des glandes annexes mâles au cours de la vie imaginale chez *Acanthoscelides obtectus Say* (coleoptera Bruchidae) C.R. Acad. Sc. 280: 889–892.
Huignard, J. & Biemont, J.C., 1979. Vitellogenesis in *Acanthoscelides obtectus* Say (coleoptera Bruchidae). II: The conditions of vitellogenesis in a strain from Columbia. Comparative study and adaptative significance. Int. J. Invert. Reprod. 1: 233–244.
Ibrahim, A.M. & Coyne, D.P., 1975. Genetics of stigma shape, cotyledon position and flower color in reciprocal crosses between *Phaseolus vulgaris* and *Phaseolus coccineus* and implications in breeding. J. Amer. Soc. Hort. Sc. 100: 622–624.
Kaplan, L., 1956. The cultivated beans of the prehistoric southwest. Ann. Mo. Bot. Gard. 43: 189–251.
Kaplan, L., 1965. Archeology and domestication in American *Phaseolus* (beans) Econ. Bot. 19: 358–368.
Kaplan, L., Lynch, T.F. & Smith, C.E., 1973. Early cultivated beans (*Phaseolus vulgaris*) from an intermontane Peruvian valley. Science 179: 76–77.
Labeyrie, V. & Maison, P., 1954. Sur les relations entre la ponte d'*Acanthoscelides obtectus Say* dans la nature et les stades phenologiques de *Phaseolus vulgaris L.* C.R. Acad. Sc. 238: 1920–1922.
Lack, D., 1965. Evolutionary ecology. J. Applied. Ecol., 223–231.
Leroi, B. & Jarry, M., Etude comparée de la ponte d'*Acanthoscelides obtectus* sur graines et gousses de six espèces de Phaseolus et des possibilités de développement dans ces graines (in preparation).
Masaki, S., 1980. Summer diapause. Ann. Rev. Entomol. 25: 1–25.
Miranda Colin, S., 1966. Identification de las especies Mexicanas y cultivadas de genero *Phaseolus.* Escuela Nacional de Agricultura Serie de Investigaciones Chapingo. Mexico 8: 1–15.
Miranda Colin, S., 1967. Infiltracion genetica entre *Phaseolus vulgaris* y *Phaseolus coccineus.* Sc. Invest. 9: 48 pp.
Miranda Colin, S., 1968. Origen de *Phaseolus vulgaris L.* (Frijol comun) Agronomia tropical, XVIII: 191–205.
Morton, E.S., 1971. Nest predation affecting the breeding season of the clay colored Robin, a tropical song bird. Science 171: 920–921.
Ouchterlony, O., (1949). Antigen antibody reactions in gels. Acta Pathol. Microbiol. 26: 507–515.
Pajni, H.R. & Sood, S., 1975. Effect of pea pollen feeding in naturation and copulation in the beetle *Bruchus pisorum L.* Ind. J. Exper. Biol. 13: 202–203.

Prevett, P.F., 1967. (a) Notes on the biology, food plants and distribution of Nigerian Bruchidae (Coleoptera) with particular reference to the northern region. Bull. Ent. Soc. Nigeria. I: 3–6.
Prevett, P.F., 1967. (b) Observation on the biology of six species of Bruchidae (Coleoptera) in northern Nigeria. Entomologists monthly magazine 102: 174–180.
Servicio Meteorologico National 1976. Normales climatologicas. Secretaria de Agricultura y Gonaderia Ed.
Smith, C.E., 1968. The new world centers of origin of cultivated plants and the archeological evidence. Econ. Bot. 22: 253–266.
Steffan, J.R., 1980. Etat actuel des connaissances sur les entomophages attaquant les Bruches. Ecology of Bruchids attacking legumes (pulses). Junk, Ed.
Tauber, M.J. & Tauber, C.A., 1976. Insect seasonality: diapause maintenance, termination, and post diapause development. Ann. Rev. Entomol. 21: 81–107.
Watanabe, T.R. & Kawanishi, M., 1979. Mating preference and the direction of evolution in Drosophila. Science 205: 906–907.
1979. Tropical legumes: resources for the future. Nat. Acad. Sc. 331 pp.

Authors' addresses:
Institut de Biocenotique Expérimentale des Agrosystèmes (ERA CNRS 328)
Université François Rabelais
Parc de Grandmont
37200 Tours
France

Instituto de Ecologia
Nuevo Bosque de Chapultepec
Apartado Postal 18–845
Mexico 18 D.F.,
Mexique

SOME ECOLOGICAL FACTORS AFFECTING DIAPAUSE IN ADULTS OF *ACANTHOSCELIDES OBTECTUS* FROM MEXICAN MOUNTAINS

I. HODEK, A. BONET & M. HODKOVÁ
(*Tours/Prague/Mexico*)

INTRODUCTION

In the course of the study of *Acanthoscelides obtectus* Say (Bruchidae: Col.) in the countries of its presumed origin, certain peculiarities of the life-cycle were found. Thus in the Colombian populations from higher altitudes important prolongation of the pre-oviposition period was ascertained (Huignard & Biemont, 1978) and in a Mexican population, originating from Tepoztlan, features indicating the adult diapause were observed. (For details on habitat and host-plants see Biemont & Bonet, this volume.)

MATERIAL AND METHODS

Sampling

The second author collected three samples in Tepoztlan (50 km south from Ciudad Mexico, state Morelos, at 1900 m altitude) in collaboration with: M. Jarry (late winter 1979), J.C. Biemont (summer 1979) and J. Pouzat (late winter 1980). Whereas the winter samples emerged in the laboratory from the previously collected pods and grains, in summer the beetles were collected in rolled dead leaves hanging from the plants.

Breeding conditions

The offsprings (F_1-gen.) of the summer sample were bred under 5 conditions (Table 1) to attempt the prevention of diapause. In all rearings, the beetles were supplied with water, mixture of pollen with honey (collected by bees) and beans (*Phaseolus vulgaris*). The pollen was exchanged once per week in lower humidity and twice per week in higher humidity. In the containers (13 x 13 x 7 cm) with lids with holes (covered with nylon) the rel. humidity ranged between 50–70%; in some experiments about 100% r.h. was produced by using solid lids and adding inside an open container with moist cotton.

Table 1. Experimental conditions.

No	Temperature (°C)	Photoperiod
1	25°/15°	12 L:12 D
2	21°/13°	12 L:12 D
3	26°/18°	16 L: 8 D
4	20°/15°	18 L: 6 D
5	19–13 hr: 20°–22° 13–19 hr: 35°–38°	natural

Piercing of pods and providing shelters

The bean pods used were dry and hard, and evidently could not be pierced by the females for oviposition. As it was observed that the females were stimulated to oviposition by small incidental cracks in the pods, the pods were artificially pierced by a pin (15 holes per pod).

In the nature the beetles estivate sheltered in dead wrinkled leaves, still suspended on trees (Biemont & Bonet, this volume). The possible effect of such hiding places was studied in some experiments. Artificial shelters were represented by rolled paper with 0.5 mm diameter.

Storage

Moderately low temperature (14°–15°) and paper shelters described above were used in the course of diapause development to decrease mortality.

RESULTS AND DISCUSSION

Year cycle of abiotic conditions, host plant and insect

The presumed life-cycle of the Tepoztlan populations of *A. obtectus* is depicted in the relation to the principal two abiotic factors which may influence it – precipitation and photoperiod (Fig. 1). The yearly alternations of temperature are not very important: the highest aver. maxima are around 30° in March, April and May and the lowest aver. maxima of 25.5° occur from September to January. The analogous minimal values are 17° and 13°, resp.

The life-cycle of *A. obtectus* is univoltine in Tepoztlan, evidently due to the host-plant. The females apparently oviposit into the mature pods in November or December and the pre-imaginal development continues in January to March. The adults which emerge in spring (or at least most of them) do not reproduce: they enter the imaginal diapause, which bridges the whole humid period lasting about 6–7 months. Thus the estival diapause of Tepoztlan populations of *A. obtectus* differs substantially from the other described cases of tropical estivation, which – on the contrary – bridge the dry period, while the development occurs at the onset of the humid season. Such is the case eg. of the pupal diapause of *Diparopsis watersi* (Galichet, 1964; Jacquemard, 1976) or the larval diapause of *Chilo* spp. (Scheltes, 1978).

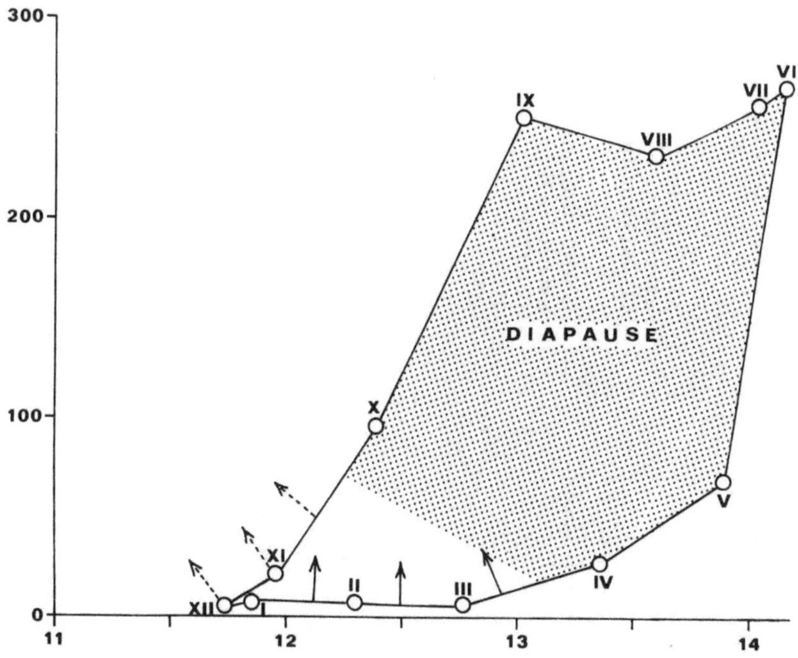

Fig. 1. Life cycle of Acanthoscelides obtectus in the Tepoztlan region, in relation to photoperiod and humidity.

Abscissa: duration of photophase (hr/24 hr); ordinate: precipitation (mm/month); roman numerals: months; arrows – dotted: oviposition, full: emergence of adults.

Activation of summer 1979 sample

The principal aim of the first activation experiments (1979) was to get descendants for crossing and other experiments; due to a small number of specimens, all were placed into presumably optimal conditions (Table 2, exp. III).

In contrast to samples from late winter this sample represented the estivating population. As the beetles were sampled in the middle of the rainy season, we assured high humidity (cca 100%) for the first 4 weeks of the experiment to simulate the continuation of natural estivation conditions. The natural photoperiod decreased in September and October more steeply (14 hr 37 min–10 hr 56 min) than in Tepoztlan (13 hr 15 min–12 hr 10 min). The beetles were very active in the warm afternoon: they were flying and feeding on pollen. The intake of food was indicated by a high amount of yellow faeces, contrasting with the brown faeces at the onset of the experiment. The much increased number of beetles on freshly pulverised supply of pollen showed its higher attractivity.

The oviposition begun after 4 weeks and it gradually increased until 3.4 eggs/female/day (Table 2, Fig. 2). The highest number found in one pod after 10 days was 389 eggs. The oviposition was still quite high at the end

Table 2. Reproductive activity in *Acanthoscelides obtectus* from Tepoztlan.

Experiment	I	II	III	IV	V
n at onset of experiment	6 ♀♀ + 7 ♂♂	10 ♀♀ + 9 ♂♂	26 ♀♀ + 20 ♂♂	35 ♀♀ + 35 ♂♂	35 ♀♀ + 35 ♂♂
Sample origin	pods/beans *P. vulgaris*		dead leafs on trees	pods *P. vulgaris* coll. 10 Jan	
date	Jan (emerg.)		27 Jul – 28 Aug (coll.)	11–26 Feb (emerg.)	
Storage conditions			variable (transport)	variable (transport), and from 13 Mar	
photoperiod	12 L:12 D			12 L:12 D	
temperature	25°/15°			25°/15°	
rel. humidity				100%	
Contrast conditions					
date	3–24 Aug	19 Aug–23 Sep	–	–	–
photoperiod	12 L:12 D	obscurity			
temperature	21°/13°	12° const.			
Activation conditions					
onset	24 Aug	23 Sept	13 Sept	28 Mar	
photoperiod	12 L:12 D		natural		
temperature	25°/15°		19–13 hrs: 20°–22°, 13–19 hr : 35°–38°		
rel. humidity				100%	50–70%
1st ovipos.	17 Sept	15 Oct	11 Oct	24 Apr	18 Apr
Min. duration of pre-ovipos. (period days)	24	22	28	27	20
Intensity of oviposition (eggs/♀/day)					
1st month	0–0.2	0–0.7	0–0.04	0–0.9	0–0.1
2nd month	0.2–0.5	0.8–1.7	0.6–2.7	0.2–0.3	0–0.1
3rd month	0.2–0.4	1.5–1.7	2.8–3.4	0–0.05	0.4–(1.9)

of the experiment in mid December when 12 females were still living. Although on October 30 the males were taken (for crossings) from group C, the females continued laying hatchable eggs.

As the beetles were collected rather late, this experiment did not give much information about the min. obligatory duration of diapause development: we achieved the oviposition only about 10 weeks before its occurrence in nature.

Three stimulating factors acted together, presence of oviposition substrate – pods with holes, consumption of pollen (with honey), increased illumination and temperature, causing high mobility. It is difficult to evaluate their relative importance.

Activation of winter 1979 sample

The beetles emerged in late January in the laboratory from beans and pods collected in Tepoztlan in early January. They were provided with water, pollen and bean grains – but not with bean pods – in low Petri dishes (9 x 1.5 cm) and remained non reproductive till August.

Two groups from this sample were exposed to 'contrast' conditions of lower temperature and luminosity and then returned to the previous conditions (Table 2, exp. I and II). Their rearing conditions were changed by providing larger space (13 x 13 x 7 cm) and bean pods. The oviposition begun after 3–4 weeks and the intensity of oviposition was higher in the exp. II with the more important difference between the 'contrast' and the activation conditions.

A third group from this sample, left constantly in the previous conditions, remained non-reproductive until the death of the beetles in April 1980.

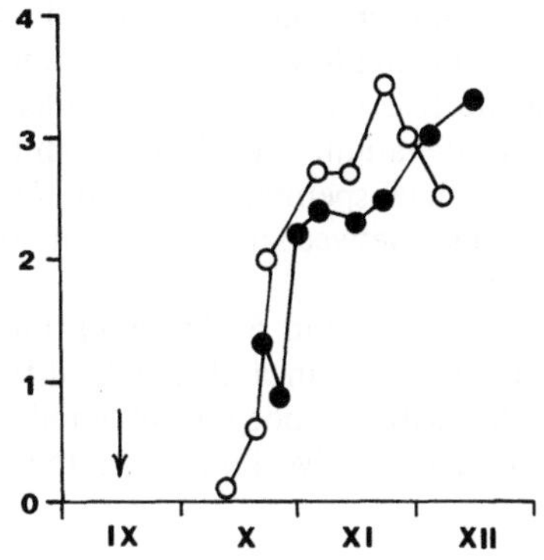

Fig. 2. Oviposition intensity in Acanthoscelides obtectus (Tepoztlan, summer sample; exp. cond.: Table 2, exp. III).

Abscissa: months; ordinate: intensity of oviposition (eggs/female/day); arrow: transfer to activation conditions; open circles: group B – always complete pairs; full circles: group C – after October 30, isolated females.

Activation of winter 1980 sample

The beetles emerged on February 11–26, 1980 in the laboratory from pods of wild *Phaseolus vulgaris* (Table 2, exp. IV and V) and *P. coccineus* (Table 3, Fig. 3, exp. I and II) collected in Tepoztlan January 10.

While in summer a representative sample of the diapausing population is concerned, this is not sure in winter samples. It cannot be excluded a priori, that a certain proportion of the emerged adults would have another behaviour than to hide into the leaves.

The possibility that some beetles can reproduce without diapause is indicated by an early oviposition (after 2–3 weeks) in the sample from *P. coccineus* (Fig. 3) in both conditions (Table 1: exp. I – cond. 5, exp. II – cond. 1). To follow the incidence of oviposition, the beetles were isolated to individual pairs in early April. Almost 50% of females begun to oviposit (Fig. 3B) and oviposited until the end of the experiment, 70 days after the onset of activation, when the dissection revealed a similar picture (Table 3). The intensity of oviposition remained, however, always low (Fig. 3A).

The average intensity of oviposition in the group rearing of the sample from *P. vulgaris* was similarly low and oviposition delay was longer – 3–4 weeks (Table 2, exp. IV and V). The oviposition was slightly more important at higher humidity, but the third month the picture was reversed.

In early July, the lower humidity replicate was discarded due to high mortality. Under high humidity the mortality was lower, but the remaining 14 females (with 18 males) discontinued oviposition. They were divided into 2 groups; one remained under high humidity, the other was exposed to lower humidity. Neither of the groups renewed the oviposition till Sept. 10.

ATTEMPT TO PREVENT DIAPAUSE IN F_1-GENERATION

The progeny of the summer sample was reared under five available conditions (Table 1), in attempt to prevent the diapause. Dissection of 102 females reared under these different conditions for 9–12 weeks revealed no important difference among the five replicates: the incidence of females without any trace of oocytes in vitellarium was always high (80–95%). All females with vitellogenic oocytes had spermatophores in *bursa copulatrix.* There was, however, no coincidence between maturation of oocytes and the decrease in the amount of fat body.

Under identical conditions (1 and 5, Table 1), the beetles of the winter 1980 sample (Table 3) responded in a strikingly different way: there were only 40–60% of females without oocytes in vitellarium. The reasons for the higher incidence of reproductive females in the winter sample will be discussed later.

Effect of pods and shelters in F_1-generation

In the course of the activation of the summer sample, we observed a different response of beetles to grains and to intact and punctured pods. In one replicate

Table 3. Dissection of females of *Acanthoscelides obtectus* from Tepoztlan (winter 1980 sample) reared for 70 days in pairs (see also Fig. 3).

Exper.	n	Ovarioles: without oocytes	oocytes in vitellarium	oocytes also in lat. oviducts
I	15	9	2	4
II	16	6	4	6

Exper.	n	Bursa copulatrix: without spermatophore	with spermatophore	Alimentary canal: empty	full
I	15	14	1	6	9
II	16	9	7	0	16

Exper.	n	Fat body: –	+	++	+++
I	15	7	2	0	6
II	16	6	2	2	6

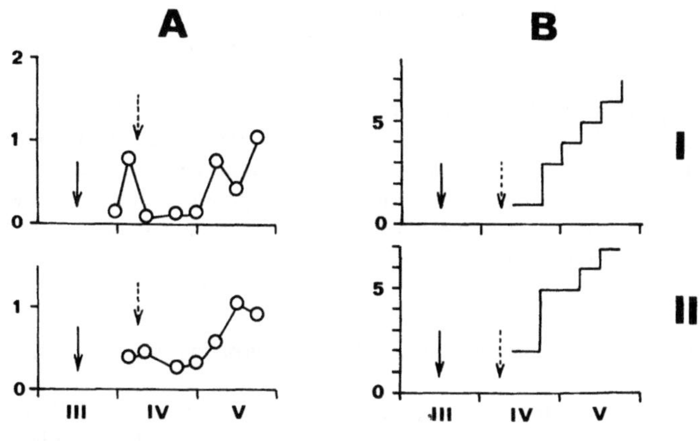

Fig. 3. Reproductive activity in Acanthoscelides obtectus (Tepoztlan, winter 1980 sample; dissection: Table 3).

Abscissa: months; ordinate: A – intensity of oviposition (eggs/female/day), non-ovipositing females included; B – accumulated number of ovipositing females; arrows – full: onset of activation, dotted: isolation of pairs.

Experiments started with 20 pairs.

of F_1-rearing the presence of shelters diminished the activity of beetles evidenced by smaller amount of yellow faeces and by less eggs.

These indications were checked in an experiment with 6 replicates (Table

Table 4. Effect of host-plant on reproduction in *Acanthoscelides obtectus* from Tepoztlan (F_1-gen.).

Experimental conditions	Pods		
	punctured	punctured	'intact'
		Shelter[1]	
	absent	provided	absent
Date of first oviposition	21 Apr	21 Apr	21 Apr
Min. duration of pre-ovipos. period (days)	16	16	16
Intensity of oviposition (eggs/♀/day):			
1st month	0–0.2	0–0.03	0–0.2
2nd month	0.2–0.3	0.01–0.05	0–0.4

[1] See methods.

Each experimental group started with 25 females and 25 males 5 Apr.

Conditions: 5 (Table 1).

The females kept without pods in three other experimental groups (with intact beans, polished beans, or without beans) did not oviposit at all.

4). Beetles developed at conditions 1 (Table 1), emerged 27 December–7 January, and after the storage in 'artifical diapause' for 13 weeks at 14°–15°, 50–70% rel. humidity and darkness they were exposed to conditions 5 (Table 1).

The females oviposited only in the three replicates with bean pods. The presence of shelters markedly decreased the average intensity of oviposition to about one tenth during both months of experiments. No difference was found between punctured and 'intact' pods, apparently due to the fact, that the Mexican pods used sometimes slightly cracked, so that a tiny fissure appeared near the petiolus.

Effect of relative humidity and diapause duration in F_1-generation

The females of winter 1979 sample did not oviposit in small Petri dishes, where the moist cotton produced ca. 100% rel. humidity. Under natural conditions, the beetles diapause during the humid season (Fig. 1), while the oviposition occurs in the dry season. We checked the assumption that very high relative humidity directly inhibits the reproductive activity of *A. obtectus* from Tepoztlan.

Another hypothesis checked in this series of experiments was the assumption that the diapause intensity and/or incidence decreases with time within 5 months. Two samples were therefore tested in every rel. humidity: the beetles developed under conditions 1 (Table 1) and stored 10 or 19 weeks in 'artificial diapause' at 14°–15°, 50–70% rel. humidity and darkness.

Table 5. Reproductive activity in *Acanthoscelides obtectus* from Tepoztlan (F_1-gen.) under two relative humidity conditions and after different duration of artificial diapause (for dissection see Table 6).

Experiment	0	I	II	III	IV
Date of adult emergence	13–17 Dec	7–15 Jan		20–23 Dec	
Exposure to 14°–15° (weeks)	0	10–11	10–11	19	19
Date of transfer to activation conditions with	17 Dec	26 Mar	26 Mar	30 Apr	30 Apr
rel. humidity (ca.%)	50–70	100	40–50	100	40–50
Date of first oviposition	9 Feb	10 Apr	21 Apr	14 May	–
Min. duration of pre-ovipos. period (days)	54	15	26	14	–
Intensity of oviposition (eggs/♀/day)					
1st month	0	0–0.7	0–0.2	0–0.1	0
2nd month	0–0.1	0.5–0.8	0.1	0.2–0.9	0

Each experimental group started with 20 females and 20 males at conditions 5 (Table 1).

No clear indication of diapause development was found. In the low humidity replicates, the long storage sample did not reproduce at all (Table 5, exp. IV). In the higher humidity, the min. pre-oviposition period and the intensity of oviposition were almost the same after both durations of 'artificial diapause' (Table 5, exp. I and III) and the dissection showed only a slightly lower incidence of non-reproductive females (43% compared with 57%) after the long storage (Table 6, exp. I and III). A certain possibility of progress of diapause development within the first 10 weeks was indicated by the much longer duration of pre-oviposition period in the adults activated immediately after emergence (Table 5, exp. 0); they included, however, the same incidence of non-reproductive females as the beetles after diapause (Table 6, exp. 0).

The effect of the relative humidity was rather opposite to our assumption. Under high rel. humidity we found shorter min. pre-oviposition period, higher aver. intensity of oviposition (Table 5) and lower incidence of non-reproductive females: 57% and 43% in comparison with 84% and 100% under lower rel. humidity (Table 6). It should be stressed that the beetles were stored in the laboratory under lower humidity and the humidity prerequisite for diapause development – if there is some – could remained unsatisfied. Also the control humidity in the course of activation was perhaps too low.

In a still running experiment with beetles stored in 'diapause' for 6 months we are obtaining in the course of August and September a reversed result under conditions 5 (Table 1): no oviposition under ca. 100% rel. humidity and oviposition under 50–70%.

CONCLUSIONS AND DISCUSSION OF LITERATURE

Oviposition substrate (bean pods)

Our preliminary experiments proved unambiguously, that – in contrast to the European strain – bean pods are a prerequisite for the oviposition of the Tepoztlan populations of *A. obtectus.* Very probably the beetles oviposit into mature but still not hard dry pods, in which they are able to pierce holes. They may oviposit also into crevices in cracked pods. In our experiments, man-made holes were a substitution which enabled the oviposition into too hard pods. The summer diapause is supposed to be terminated by specific stimuli (Tauber & Tauber, 1976; Masaki, 1980). If the summer diapause of *A. obtectus* adults from Tepoztlan is really terminated by the availability of oviposition substrate, i.e. the bean pods of appropriate maturity, it would be the first case reported.

Adult food (pollen)

In all experiments pollen (with honey) was supplied and the adults consumed it eagerly. The absence of pollen (or at least of the pollen of the 'right' plants) at the end of the dry season may induce diapause and/or its presence in late rainy period may terminate it. The estivating adults were found to be mobile in August (Biemont & Bonet, this volume) and can be attracted to pollen during their periods of activity.

The adult diapause may be induced by alimentary signals; this was supposed by Hagen (1962) for some Coccinellidae of the genus *Hippodamia*, and it was proved for *Semiadalia undecimnotata* by Iperti and Hodek (1974) and by Rolley *et al.* (1974). In many phytophagous insects the induction of diapause is favoured by feeding on matured host-plants (see references in the last mentioned paper and in the review by Masaki (1980). In the *mohave* strain of *Chrysopa carnea*, the summer diapause was reported to be a response to starvation, while presence of food terminates it (Tauber & Tauber, 1973).

Humidity

We did not prove experimentally the necessity of a specific relative humidity in the course of diapause development. The beetles stored at 14–15° and 50–70% relative humidity showed much lower reproductive activity, even after 19 weeks of 'artificial diapause', than the estivating beetles, sampled in the nature at about the same age. This difference may indicate that some prerequisite for diapause development was not fulfilled.

In experiments, where very high relative humidity (ca. 100%) was compared with very low humidity (ca. 40–50%), the high humidity appeared more favourable in beetles stored for 10–19 weeks in 'artificial diapause'. In another experiment the 'normal' humidity of about 50–70% was more suitable than ca. 100% for beetles stored 6–7 months. This might indicate the necessity of high humidity during diapause, but of low humidity for the post-diapause development.

Table 6. Reproductive activity of *Acanthoscelides obtectus* from Tepoztlan (F_1-gen.) under two relative humidity conditions and after different duration of artifical diapause.

Exper.	n	Relative humidity (%)	Duration of diapause (weeks)	Ovarioles			Bursa copulatrix		Fat body				Alim. canal	
				without oocytes	oocytes in vitellarium	oocytes also in lat. oviducts	without spermatophore	with spermatophore	–	+	++	+++	empty	full
0	18	50–70	0	15	3	0	15	3	12	4	2	0	–	–
I	17	100	10	10	5	2	12	5	0	3	0	14	0	17
II	19	40–50	10	16	2	1	17	2	0	1	0	18	1	18
III	14	100	20	6	7	1	8	6	0	3	3	8	5	9
IV	17	40–50	20	17	0	0	17	0	0	8	9	0	5	12

Females dissected 97 days (exp. 0), 41 days (exp. I and II), or 44 days (exp. III and IV) after the transfer to activation conditions 5 (Table 1). For other parameters of reproductive activity see Table 5.

Photoperiod

For technical reasons no experiments were made to study the effect of photoperiod on the diapause of *A. obtectus* which, however, may be presumed. Since the work on *Nomadacris septemfasciata* (Norris, 1965) it is well known that even slight changes in photoperiod in low latitudes may serve as a signal for the seasonal changes in availability of host plants and food. Thus the difference in daylength between the winter and summer solstice, which is more than $2\frac{1}{2}$ hours in Tepotzlan (Fig. 1), would be highly sufficient for the regulation of year cycle of *A. obtectus.*

Our good activation achievements with the summer sample were obtained under natural decrease of photoperiod in autumn, i.e. in the period when the females oviposit in Tepoztlan.

In some cases, e.g. in the *Chrysomelid Galeruca tanaceti* (Siew, 1966), it was reported that the estival diapause is terminated by decreasing short day. The *Staphylinid Lathrimaeum atrocephalum* needs short day after estivation to oviposit (Topp, 1979).

Polymorphism

In several experiments we obtained the oviposition in adults relatively early after their emergence. Although the incidence of ovipositing females was often quite low, this finding indicates a heterogeneity of the Tepoztlan population of *A. obtectus.* The fact that the progeny of the summer sample responded in a different way than the 1980 winter sample may indicate that even in Tepoztlan the adults emerging in late winter and early spring do not behave all in the same way. It cannot be a priori excluded that the beetles, diapausing in summer in wrinkled leaves, represent only a proportion – although perhaps large – of the total population.

The Tepoztlan population of *A. obtectus* is possibly polymorphic as far as diapause is concerned, similarly to *Coccinella septempunctata* (Hodek and Růžička, 1979) or *Aelia acuminata* (Hodek and Honěk, 1970). An important polymorphism in the oviposition response to stimuli of the host-plant was discovered in the French populations of *A. obtectus* (Labeyrie, 1968).

REFERENCES

Galichet, P.F., 1964. Diparopsis watersi, Lepidoptera, Noctuidae, ravageur du cotonnier en Afrique centrale. Cot. Fib. Trop. 19: 437–518.

Hagen, K.S., 1962. Biology and ecology of predaceous Coccinellidae. Ann. Rev. Entomol. 7: 289–326.

Hodek, I. & Honěk, A., 1970. Incidence of diapause in Aelia acuminata (L.) populations from southwest Slovakia (Heteroptera). Věst. Čs. zool. Spol. 34: 170–183.

Hodek, I. & Růžička, Z., 1979. Photoperiodic response in relation to diapause in Coccinella septempunctata (Coleoptera). Acta ent. bohemoslov. 76: 209–218.

Huignard, J. & Biemont, J.C., 1978. Comparison of four populations of Acanthoscelides obtectus (Coleoptera: Bruchidae) from different Colombian ecosystems. Assay of interpretation. Oecologia (Berl.) 35: 307–318.

Iperti, G. & Hodek, I., 1974. Induction alimentaire de la dormance imaginale chez

Semiadalia undecimnotata Schn. (Col. Coccinellidae) pour aider à la conservation des Coccinelles élevées au laboratoire avant une utilisation ultérieure. Ann. Zool. – Écol. anim. 6: 41–51.

Jacquemard, P., 1976. La diapause de Diparopsis watersi (Roths.)(Lepidoptera Noctuidae) dans le nord du Cameroun. Cot. Fib. Trop. 31: 297–311.

Labeyrie, V., 1968. Longévité et capacité reproductrice de lignées d'Acanthoscelides obtectus sélectionnées en fonction de la réponse aux stimuli de ponte. C.R. Soc. Biol. 162: 2203–2206.

Masaki, S., 1980. Summer diapause. Ann. Rev. Entomol. 25, 1–25.

Norris, M.J., 1965. The influence of constant and changing photoperiods on imaginal diapause in the red locust (Nomadacris septemfasciata Serv.). J. Insect Physiol. 11: 1105–1119.

Rolley, F., Iperti, G. & Hodek, I., 1974. Influence de la nourriture aphidienne (selon l'age de la plante-hôte à partir de laquelle les pucerons se multiplient) sur l'induction de la dormance chez Semiadalia undecimnotata SCHN. (Col. Coccinellidae). Ann. Zool. – Écol. anim. 6: 53–60.

Tauber, M.J. & Tauber, C.A., 1973. Nutritional and photoperiodic control of seasonal reproductive cycle in Chrysopa mohave. J. Insect Physiol. 19: 729–736.

Tauber, M.J. & Tauber, C., 1976. Insect seasonality: diapause maintenance, termination, and postdiapause development. Ann. Rev. Entomol. 21: 81–107.

Topp, W., 1979. Vergleichende Dormanzuntersuchungen an Staphyliniden (Coleoptera). Zool. Jb. Syst. 106: 1–49.

Scheltes, P., 1978. The condition of the host plant during aestivation-diapause of the stalk borers Chilo partellus and Chilo orichalcociliella (Lepidoptera, Pyralidae) in Kenya. Ent. exp. & appl. 24: 679–688 (Proc. 4th Insect/Host Plant Symp.).

Siew, Y.C., 1966. Some physiological aspects of adult reproductive diapause in Galeruca tanaceti (L.). Trans. R. ent. Soc. London 118: 359–374.

Authors' addresses:
I. Hodek and M. Hodková:
Inst. Entomology,
Czechoslovak Academy of Sciences,
12800 Prague
Czechoslovakia

A. Bonet:
Inst. Ecologia,
Nuevo Bosque de Chapultepec,
Mexico 18,
D.F. Mexico

IMPORTANCE OF ALLELOCHEMICS IN PLANT INSECT RELATIONS

Personal reminiscences

GOTTFRIED FRAENKEL
(*Urbana, Ill., U.S.A.*)

I have been given the task to give an introductory lecture about the function of the secondary plant substances (allelochemicals) in the coevolution of plants and insects. This I consider neither possible nor necessary in the context of this symposium which deals with a particular aspect of this problem, that of the specificity of bruchid beetles on leguminous seeds. Ever since I introduced the concept of the biological role of these substances (1959), there have been about a dozen books dealing largely or entirely with this subject, about 10 symposia, and literally hundreds of individuals articles. It is entirely beyond my capabilities to give a comprehensive account of this subject at this occasion.

This audience is composed of two kinds of participants: Those who are well versed in this subject and are actively working on the relations between bruchids and their food plants, and those familiar more with control and the administrative aspects. For the former group, I propose to give a brief historical overview about how this concept of secondary plant substances and coevolution has shaped in my own mind. For the latter group, I will state in a few sentences the basic principles underlying this concept.

Talking now in part by hindsight, this concept seems to have evolved over a period of about 15 years and had several roots.

1. During the second world war, I studied the food requirements of insects from the point of view of dietetics: which chemical substances are required in the development of insects. By about 1948 it had become clear that these were basically the same for all insects, and for that matter, with certain exceptions, largely identical with what was then known for vertebrates: proteins or rather the 10 essential amino acids, certain minerals, about 10 B-vitamins and, required only by insects, sterols. On the other hand, carbohydrate requirements were very variable in different organisms, qualitatively and quantitatively.

2. During the same time, in war-time England, I participated in a study committee, organized by the Fabian Society, then as now a research organization associated with the British Labour Party, which concerned itself with the problems of British agriculture, both from the point of view of war-time conditions, and outlook for the future. The results came out in a book:

Towards a socialist agriculture (F.W. Bateson, ed., London, Victor Gollancz, 1946), which contained a chapter by me: *Britain's nutritional requirements.* In this chapter, great emphasis was placed on the dietary importance of green and yellow vegetables, mainly because of their content of vitamins A and C, and their protein content. Little was known then about their content of the B-vitamins, and the concept of first class proteins, of which leaf proteins are a good example, hardly existed at the time. This certainly prepared me to look at leaves as a source of food.

3. By what I consider now a happy coincidence, at the same time a young Canadian entomologist, A.J. Thorsteinson, turned up in my laboratory and embarked on a thesis about the role of mustard oil glucosides in the food selection of the cabbage diamond moth, *Plutella maculipennis.* This was started as a reinvestigation of a classical paper by Verschaeffelt (1910) on a similar topic, using the Pieris white butterfly. Verschaeffelt's very fundamental results, probably the first conclusive proof of a group of secondary plant substances being entirely essential for the feeding of an insect, had never been reinvestigated and the choice of another cabbage insect, *Plutella,* proved a very happy one. Thorsteinson's results entirely corroborated Verschaeffelt's, and in many respects, experimentally and conceptually, extended them very considerably. This work was finished by 1947, but publication was considerably delayed (Thorsteinson, 1953).

4. In 1951, the first of many subsequent symposia on insects and plants was held at the International Congress of Entomology at Amsterdam. In my contribution, I made it clear that the food requirements of all insects were basically very similar, and that the leaves of plants contained all the required nutrients in adequate amounts. This posed the question of why do not *all* insects develop on *all* plants. It was here tentatively suggested that it was the presence of 'odd' chemical substances in which the plants differ, what we would now call secondary plant substances or allelochemicals, which rendered a plant a suitable or unsuitable food.

5. These ideas were further elaborated and documented in a lecture given at the International Congress of Zoology at Copenhagen in 1953, which in nucleo contained the basic ideas of the concept of coevolution of insects and plants, 'the reciprocal adaptive evolution which occurred in the feeding habits of insects and the biochemical characteristics of plants'. A number of proved or hypothetical examples were quoted (Fraenkel, 1956 (1953)).

6. During the following years, I gave numerous lectures on this topic, and embarked, with several doctoral candidates, on a series of studies involving the silkworm, *Bombyx mori*; the Southern army worm, *Prodenia eridania*; the Colorado potato beetle *Leptinotarsa decemlineata*; the tobacco hornworm, *Manduca sexta,* and other insects. But it was not until 1959 when the comprehensive article on the 'Raison d'être of the secondary plant substances' appeared in *Science*, and virtually in the same form in the *Transactions of the International Congress of Biochemistry* in Vienna in 1959, where it had been given as a lecture in a symposium on Insect Biochemistry (Fraenkel, 1959a and b).

The views expressed in these articles found relatively little, and at that often negative response (cf. Fraenkel, 1970), until the very important article on coevolution in butterflies and plants by Ehrlich & Raven (1964), but by about 1969, the field had broken wide open, with a symposium about every year and a flood of hundreds of papers, continuing unabatedly until today.

For those here not at all familiar with these concepts, I shall try to state the essential features in a nutshell. The mode of selection by insects of their hostplants seems very haphazard, some feeding on many, but by no means all plants species (polyphagy), others on relatively few (oligophagy), and other on only one or a few very closely related species (monophagy). All leaves contain more or less the same nutrients, including all those which all insects require. If nutritionally all leaves are equal, why that specific selection? The answer is the presence in different plants of many different substances, alkaloids, essential oils, glucosides, tannins, etc., which give different plant families, and different individual plants, their chemical characteristics.

These 'secondary' substances have arisen in evolution as a defense mechanism of plants against insects (and other organisms). Insects in turn developed defense mechanisms against such substances by which they became able to deal with them, in many different ways, different insects in relation to different substances. In many cases, this led further to a situation where insects became attracted by such substances and would not feed in their absence. This is the case with monophagous and oligophagous insects, while polyphagous insects feed on all plants which do not contain substances repellent to them. This accounts for the immense variety of secondary substances and of insect-hostplant relationships. This also accounts for the fact that, with few exceptions, no physiological functions of such compounds in the metabolism of plants are known.

This is essentially the view which was stated in the 1959 papers and which, with little modification, appears to be generally accepted today.

The bruchid beetles and their food, the leguminous seeds, represent good examples of this insect-hostplant relationship.

REFERENCES

Ehrlich, P.R. & Raven, P.H., 1964. Butterflies and plants: A study in coevolution. Evolution 18: 586–608.

Fraenkel, G., 1951 (1953). The nutritional value of green plants for insects. Trans. IXth International Congress of Entomology, Amsterdam, vol. 2, 90–100.

Fraenkel, G., 1956. Insects and plant biochemistry. The specificity of food plants for insects. Proc. XIV International Congress of Zoology, Copenhagen, 5-12, August 1953, 383–387.

Fraenkel, G., 1959. The raison d'être of secondary plant substances. Science 129: 1466–1470.

Fraenkel, G., 1959. The chemistry of host specificity of phytophagous insects. Fourth Intern. Congr. of Biochemistry. vol. XII. Biochemistry of Insects. Pergamon Press, London, 1–14.

Fraenkel, G., 1970. Evaluation of our thoughts on secondary plants substances. Ent. exp. and appl. 12: 473–486.

Thorsteinson, A.J., 1953. The chemotactic responses that determine host specificity in an oligophagous insect. (*Plutella maculipennis* Curt., Lepidoptera). Canad. Zool. 31: 52–72.

Verschaeffelt, E., 1910. The cause determining the selection of food in some herbivorous insects. Proc. Acad. Sci. Amsterdam 13: 536–542.

Author's address:
Dept. of Entomology
University of Illinois
Urbana, Illinois 61801
U.S.A.

THE ROLE OF SENSE ORGANS IN THE RELATIONS BETWEEN BRUCHIDS AND THEIR HOST PLANTS

J. POUZAT
(Tours, France)

INTRODUCTION

The first level of reaction of animals towards the environment's complexity (diversity of its parameters, temporal variations, accidental perturbations) is behavioural (Slobodkin, 1968). The second level of reaction is physiological (vegetative reactions in the broadest sense). These two types of reactions occur at the level of the individual, whereas a third one takes place at the population level and depends on the population's gene pool: polymorphism enables populations to adapt to different environments and variations therein. These in turn will modulate the genetic composition of populations.

These ideas are illustrated in the following brief review of the rôle of sensory organs in the Bruchids.

Behavioural analysis is obviously a key step in the study of a species whether we are interested in fundamental biology or in practical applications. If the aim is to reduce crop losses due to Bruchids at an economically acceptable level, some knowledge of their behaviour is necessary. If we provisionally accept that Bruchids are stored product insects (the famous so-called 'post-harvest' category), it remains that infestation can only occur before or after harvesting. Since in fact, the main infestation occurs before harvest, crop protection requires more subtle methods (than fumigating a silo). Therefore information on Bruchid behaviour is necessary.

Behavioural studies begin with a descriptive stage and then proceed with causal analysis. The first observations are done in nature. Thereafter, there are always strong interactions between behavioural observations (whether in the laboratory or in nature) and experiments. A sound knowledge of the natural cycle is necessary but there is obviously much information which cannot be collected on natural populations in the field. In a given natural situation, a female's reaction may be influenced by insemination, or by the fact that she has begun to lay eggs, etc. Some of these uncertainties can be clarified by close examination of a sample of the population under study. We can thus know the percentage of mated females. However, many other features, obviously remain unknown since this requires knowledge of the previous history of each female. Studies in the controlled conditions of the

Series Entomologica, Vol. 19: ed. by V. Labeyrie

laboratory are unescapable. This is also the only way of gaining some idea of the extent of polymorphism in the population. But breeding animals and experimenting necessarily involves simplification of the 'milieu'. Results obtained should therefore not be too quickly transposed. Furthermore, the design of experiments should always be made from an ecological point of view, i.e., the factors we intend to study should always be chosen according to what has been observed, or what is assumed to occur in nature.

This study of behaviour obviously implies some examination of the functioning of sense organs. In a recent paper Autrum (1977) defined behaviour as the touchstone of sensory physiology: 'First, observation shows us forms of behaviour that require explanation. Second, the results of the physiological, physical and biochemical analysis of particular processes must ultimately be testable – and tested – against the behaviour'.

Moreover, it should be emphasized that what is observed in a given experiment is not fully explainable by mere reference to the factors present at that point in time. Reducing explanations to the S-R scheme can be very dangerous. The importance of the animal's history has already been stressed. Generally speaking, it is necessary to take into account what Wigglesworth (1965) called 'third variables', among which should be noticed: various rhythms, memory, 'autonomous' central patterning of motor acts, exact functional relations between sensory organs and C.N.S., etc.

Besides their rôle in a given behaviour pattern, sensory factors may also have delayed actions on other functions, mainly vegetative ones (releaser effect and primer effects, respectively (Wilson & Bossert, 1963). The study of sensory organs should therefore not be confined to their rôle in behaviour (and memory) but should also include these primer effects. These were formerly limited to the effects of pheromones, but many other stimuli may have a primer effect.

MAIN STEPS IN BRUCHID LIFE CYCLE

After emerging, adults remain for a while where larval development occurred, pods or seeds according to the species. In diapausing *Acanthoscelides obtectus* populations, delayed dehiscence of pods accounts for the more or less extended stay on the host plant. In 'domestic' populations of this species, an important proportion of the emerging adults may remain on seeds in storage and reproduce there.

It is most likely that, in this species as well as in others, take-off depends on climatic factors, mainly light and temperature. This seems established in the domestic *Acanthoscelides obtectus* population (Arnoux *et al.*, 1958) and has been experimentally studied by Perttunen et Hayrinen (1969). They describe (1970) a polymorphism in take-off activity. However, this last point does not seem convincingly established in view of the difficulties involved in the study of behavioural polymorphism. For a given individual, a behavioural sequence is not always reproducible in the same conditions, and this, independently of factors such as habituation or ageing. In many cases, 'third variables', leading to much apparent inconsistancy are usually present. This is particularly evident in behaviours such as take-off.

The following stage till 'landing' on host plant is generally poorly known in the domestic *Acanthoscelides obtectus*; in diapausing *Acanthoscelides obtectus* and *Bruchus pisorum* a more or less extended period is spent in dry curled-up leaves remaining on various plants of the ecosystem. In many species, feeding behaviour certainly occurs either on flowers (pollen, nectar) or on saprophytic fungi. However, little is known regarding what actually takes place in nature, excepted for the special case of *Bruchus pisorum* (Pajni, 1975).

The next stage is the discovery and recognition of the host-plant at least by the females. Mating occurs either before or after discovery of the host plant, the exact mating place being generally unknown. Females of the domestic *Acanthoscelides obtectus* land on the host plant, possibly after mating. They remain there for a variable length of time before beginning to lay eggs. Perception of certain chemical stimuli may have at this stage primer (delayed) effects: it has been shown experimentally (Pouzat, 1978a, b) that contact with host-plant speeds up oogenesis. In diapausing *Acanthoscelides obtectus*, these factors may contribute to ending diapause.

During this stage of identification and exploration which may last up to several days, chemical receptors (mainly taste) are involved. This is followed by a rather complicated egg-laying behaviour. Domestic *Acanthoscelides obtectus* females, when on mature pods of cultivated beans, actively look for the position along the ventral suture which they will perforate with their mandibles (Zachariae, 1958). The hole, which leads inside the pod, has been probed many times with the ovipositor during boring. After assessing with the ovipositor the state of the pod's walls, the female often lays near the funiculus, a number of eggs which may sometimes amount to several dozens. But if the pod is not mature or dry enough, egg laying does not follow perforation. The ovipositor plays an important part here, mainly because it detects tactile and moisture information coming from the hole and the rest of pod.

In every Bruchid species where this has been studied, tactile and specific chemical information always play a crucial rôle in the choice of the egg laying site and the release of eggs. The choice of the larval food base is mainly achieved by the female. The newly hatched larva shows different behaviour patterns according to the exact site where the eggs were laid. Several species glue their eggs on the seeds (*Callosobruchus maculatus, Zabrotes subfasciatus*). Their larvae bore the chorion on the side stuck to the seed and are therefore immediately in their food. With other species such as *Acanthoscelides obtectus* the situation is more complex. The newly hatched larvae actively explore the stored seeds or seeds in the pods and are therefore able to 'choose' the seed they will feed on. Furthermore Theiry (personal communication) has shown that the larvae in a pod pierce the seed rather than the pod wall in 90–95% of the cases. Choosing the correct substrate is a real problem for the larvae. This adaptive behaviour involves disciminative sensorial abilities by the larvae.

This account of the biological cycle of Bruchids will lead to a description of the rôle of sensory organs in various behaviour patterns.

VISION

No physiological study of vision has been carried out in Bruchids. Observations seem to indicate that *Acanthoscelides obtectus* females have a preference for yellow pods. (Labeyrie & Maison, 1954). The rôle of sight in take-off behaviour has already been outlined.

CHEMICAL SENSES

Olfaction

The physiology of olfaction has made great advances with studies of pheromones. Over the last few years, investigations on the rôle of olfaction in the search and discovery of host plant have yielded interesting results (Visser, 1979). In domestic *Acanthoscelides obtectus,* behavioural and electrophysiological studies have given results for two areas: Search for the male and search for host plant. Regarding the behaviour towards the males, the reactions of the females depend on age, insemination, etc. (Pouzat, 1974a, b). Reactions towards host plant are more complex. In a domestic strain with a relatively slow ovarian development, the following results were obtained with young females (less than 24 h), in a two ways choice olfactometer: 16% entered the side along which the host plant odour travels, 71% entered the other side ('clean' air, control) and 13% moved little or not. Therefore, this strain does not seem attracted by host plant odour at the beginning of its adult life. Results for another experiment, performed with 165 females of a strain with a more rapid ovarian development were as follows: 63% entering the host plant side, 5.5% the control side and 31.5% not moving. It is possible that the effect of host plant odours varies with the strain and is related with something like the 'physiological age'. It must be noticed that, in all these studies, the possible existence of an eventual polymorphism has not been examined.

From the experimental results, the host plant would have an influence at a distance on adult females. However, this question obviously requires further studies, for instance with more sophisticated olfactometer of the wind tunnel type, in which the beetle could fly and not only walk (Kennedy, 1977).

The mechanism of olfactory orientation depends on anemotaxy: young *Acanthoscelides obtectus* females walk against an odourless air current. This positive anemotaxy disappears at the age of 4–5 days. If at that age virgin females are subjected to an air stream permeated with male odour, then anemotaxy is triggered once again. Anemotactic orientation involves mechanical receptors located on antennae. This is shown by the absence of anemotaxy in very young females which had their antennae amputated, when placed in an odourless air current (their locomotory abilities remaining apparently unchanged after the operation). Receptors may either be the long sensilla located on each antennal segments or, more probably, the short sensilla located on the first segment articulating with the head capsule (Fig. 1a, b).

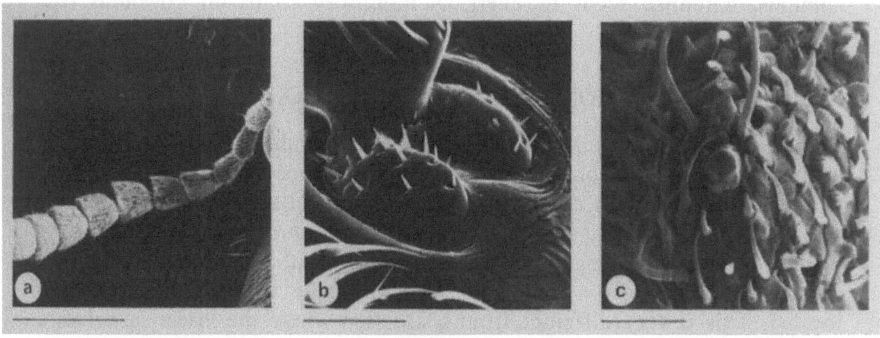

Fig. 1. Acanthoscelides obtectus. a: antenna; b: articulation of the antenna with the head; c: apex of the 11th antennal segment (scale: a: 500 μm, b: 50 μm, c: 5 μm).

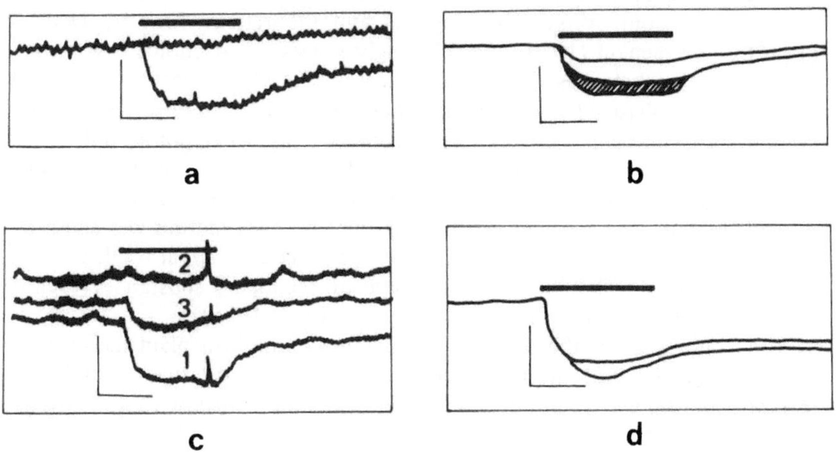

Fig. 2. EAGs of female *A. obtectus.* a: male odor (and control); b: id. showing various response amplitudes (hatched area) to the same stimulus within a 50 mn period. The upper signal is the response to the same stimulus after the antenna was subjected to ether $[(C''_2H''_5)''_2O]$ vapours; c, 1: green pod odour (bruised pod); c, 2: id. after ether vapours; c, 3: dried beans odour (the spike is an artifact); d: amyl acetate vapours. Calibration: 2s, 1 mV.

Olfactory receptors are situated on the antenna. Global electrical reaction: electroantennograms (E A G s) can be recorded on the antennae. Fig. 2 shows EAGs when male odour and host plant odours (especially green pods) are blown on the antennae. The study of olfaction should be further carried out at two levels: analysis of volatile substances of bean plants at various phenological stages and analysis of single cell responses of olfactory receptors. But these investigations are very difficult to carry out for many reasons. Almost nothing is known on the volatile substances of Leguminosae. A second reason is that Bruchids are very unsuitable for this type of work (see Fig. 1c).

Furthermore, Bruchid antennae do not only function as transducers for olfactory or mechanical stimulations: independently from any known specific stimulus, antennae seem to exert an inhibitory tonus on oogenesis (see Pouzat, 1978a).

CONTACT CHEMORECEPTION

There are many studies dealing with this sensory ability in insects, regarding behaviour as well as physiology of receptors involved in feeding. Fewer studies have dealt with egg laying. There is general agreement that contact chemoreception is a key step in the final recognition of host plant, either for feeding or for egg laying. Indeed, when a terrestrial animal perceives a non volatile substance – by contact –, he is normally where the substance is located. This is very different from a situation where the insect perceives a volatile substance somewhere.

In this context, the problem of the relative importance of secondary substances versus food substances, with regard to choice determination has caused much controversy. Experimental and theoretical work by Dethier (1954) and Fraenkel (1959, 1969) strongly suggests that choice is governed by secondary substances. This is not the topic of this paper, but it should perhaps be emphasized that this problem has passably been obscured because two levels of explanation have been confused, as Mayr points out in his paper 'Behaviour and systematics' (1958). He stresses the necessity of distinguishing: first the mechanisms by which a function is presently accomplished, second, the mechanisms through which this function was acquired during phylogeny. Adult Bruchids, which never feed on larval food, provide a very remarkable example of choice based on secondary substances (even if some species require food from the host plant (Pajni, 1975)).

The importance of the palpation behaviour on the host plant has already been emphasized and the necessity of contact with host plant established (Pouzat, 1977). Observations and ablation experiments show that maxillary palps are the main organs involved in contact recognition of host plant. Close examination of the results of ablation experiments (Pouzat, 1970, 1978a) show: (1) a very important physiological polymorphism, (2) the existence of several 'input pathways' regarding the means used to perceive a given stimulus, i.e. there is some redundancy in the picking up of information. This fact may be regarded as increasing the reliability of the insect as a biological system. Another important aspect is shown in egg laying experiments with host plant extracts. A great number of gustatory stimuli (as well as their possible interactions) and interactions between different sensory modalities namely taste and mechanoreception would be involved (Pouzat, 1976). Furthermore, the taste perception of host plant speeds up oogenesis, i.e., is a primer effect (Pouzat, 1978b).

Physiological studies of taste in Bruchids are very difficult for several reasons: (1) Almost nothing is known about specific secondary substances which may exist at the surface of host plant (the effects of pods or seeds may also be reproduced with leaves). 'Chemotaxonomy of the Leguminosae' (Harborne *et al.,* 1971) gives almost no information on this question. Furthermore, these substances are difficult to extract and to separate and are probably water insoluble.

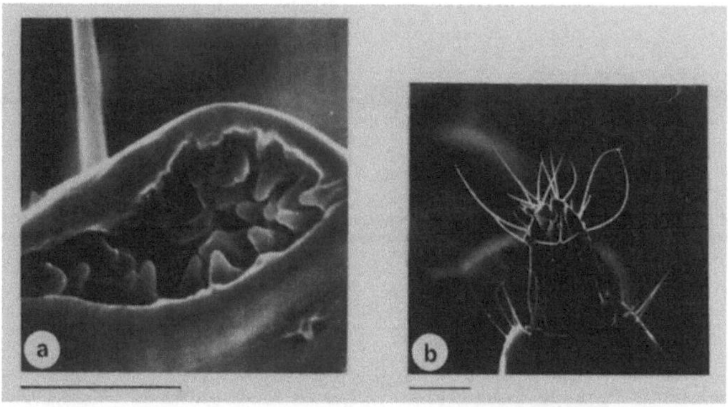

Fig. 3. A. obtectus. a: apex of maxillary palpus (shrunk owing to preparation); b: ovipositor, partly unfolded: the tip bears setae of various lengths and the first (proximal) segment bears more uniform sized sensilla. (Scale: a: 10 μm; b: 100 μm).

(2) Biological tests which may be used here (oogenesis and egg laying) are long and an electrophysiological test would therefore be valuable. Unfortunately, examination of SEM pictures of maxillary and labial palps explains the difficulties. Receptors involved in taste perception are most probably located at the apex of the palp. In *Acanthoscelides obtectus* (Fig. 3a), 40 to 50 short *sensilla basiconica* are packed on a dome 20 μm in diameter. It is not feasible to record single cell activity in these sensilla or global activity of all of them. The longest third segment of the maxillary palps is 200 μm in length.

The ovipositor is another organ on which taste receptors have been identified. Experiments with an artificial egg-laying device show that perception of liquid water by the ovipositor completely inhibited egg laying (Pouzat, 1975). Electrophysiological investigations have been carried out and it has been possible to record responses to pure water (Fig. 4a). Further studies should show if they are very specific responses. The ovipositor seems to carry 6 (or 8) taste sensilla and taste receptors do not seem to exist on the tarsi. (Fig. 3b).

It is clear that taste is a very complex sensory modality. Redundancy and interactions have already been emphasized and they constitute a very strong drawback for biological control methods intervening at this behavioural stage.

Tactile receptors

They generally do not provide very specific information: nevertheless, their rôle is very important either indirectly (antennal receptors and anemotaxy) or directly in the release of egg laying (receptors of the ovipositor). The bean weevil provides an interesting example in this instance because shape recognition is involved. Tactile receptors in this species are located at two levels of the ovipositor (Fig. 3b). The description of their part in egg laying behaviour has already been published (Pouzat, 1975). When suitably applied, tactile stimulations alone can induce egg laying.

Crevices or especially holes (similar to those it perforates in the pod), in

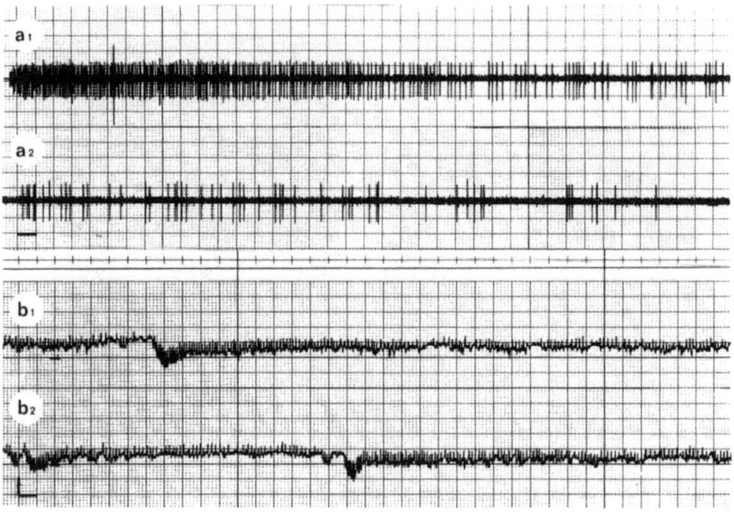

Fig. 4. A. obtectus. a1–a2: water response of a taste sensillum of the apex of the ovipositor (tip recording, A.C. coupled); b1–b2: ovipositor tactile sensillum (tip recording after cutting the sensillum's extremity, D.C. coupled): spontaneous activity and response to movements. Calibration: 1 s, 0.5 mV.

which the female is just able to introduce its ovipositor, are critical, this is why one may speak of shape recognition. Electrophysiological study of the receptors shows (Fig. 4b): (1) a tonic spontaneous activity, (2) an increased firing rate when mechanical stimuli are applied. The adaptation is feeble and under all circumstances, the firing rate is low (15–20 imp/s max.). In other bruchid species, the rôle of tactile receptors is almost unknown. With few exceptions, their egg laying behaviour is unknown. What is known in *C. maculatus* on *Z. subfasciatus*, for instance suggests that the rôle of the ovipositor is quite important but the details of its intervention are different. Egg laying behaviours and ovipositors structure are also different (Pimbert, personal communication).

POLYMORPHISM

As previously discussed, polymorphism in certain behaviours is extremely difficult to assess, especially if the behaviour under examination is brief and occurs many times during animal's life. Regarding other behaviours such as egg laying, or vegetative reactions such as oogenesis, the existence of polymorphism may be shown in a more convincing way. Two examples involving host plant will be given. The first one deals with the effects of palp amputation on egg laying. What is initially observed within the same generation (e.g. 03/1969) is a great spread of results (Fig. 5a). This shows the variation in relative importance of different taste inputs and the redundancy of input

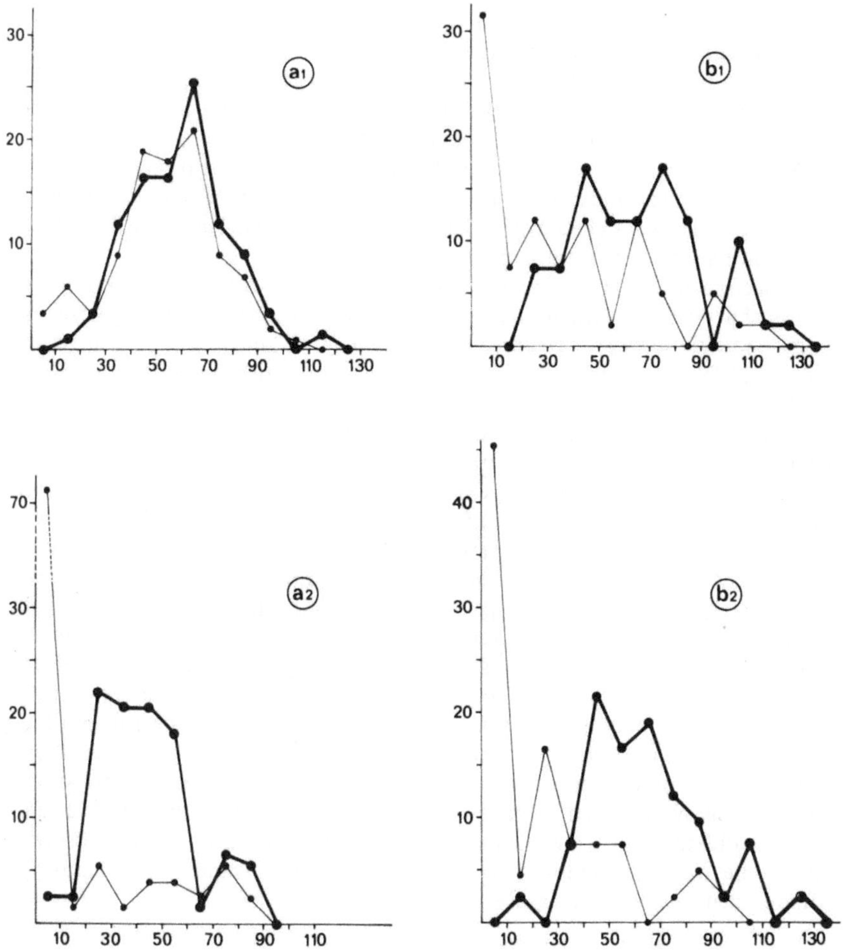

Fig. 5. Distribution of *A. obtectus* females according to the number of eggs laid (thin line) or produced (thick line) by each female. a: March 1969 generation, a1: control (unoperated) females, a2: palps removed females; b: Oct. 1969 generation, b1: palps removed females, b2: maxillary palps removed females. Abscissa: no of eggs, ordinate: no of females (absolute values) (Pouzat, 1970, 1978a).

loci. The same operation performed 6 months later on females of the same strain gave similar results. However there were important differences. Average egg laying increased from 15.25 in March to 29 in October 1969. The distribution of the females in terms of fecundity was quite different (Fig. 5b). Polymorphism is therefore very important within and between generations. Differences in the functioning of central mechanisms of integration could partly explain these variations. The contribution of taste receptors to this variability is extremely difficult to assess because of the way they function.

Dethier (1974) gave a very topical and illuminative demonstration where the same receptor, subjected successively to independent and identical stimulations did not always show the same response. These findings raise difficult questions about information encoding and the way the C N S integrates information. This problem is far from being solved.

Many other examples of polymorphism could be given regarding egg laying behaviour of *Acanthoscelides obtectus,* some of them from experiments involving non operated females. This proves that what has been described here is not a mere artifact owing to amputations.

A second example dealing with the influence of host plant on oogenesis shows that contact with host plant brings about an increase in oogenesis in virgin *Acanthoscelides obtectus*, independently of nutrition. This increase among different generations of the same strain varies between + 5% and + 80–100%. With a generation from another strain (also from a 'domestic' population) the host plant had strictly no effect under the same conditions (Pouzat, 1978a) (these experiments are done with intact females).

Thus there is practically no experiment in which polymorphism has not been observed, either within or between generations or both. Taking polymorphism into account complicates everything. This is perhaps one reason why it is so rarely mentioned in scientific papers. Experimental design e.g., is greatly complicated: at every new stage of investigation, the number of control groups increases and sometimes, several hundred females of the same generation must be available for a single experiment. A partial solution is to select strains according to certain characteristics, but this is a very tedious exercise.

Furthermore, polymorphism is a big practical challenge. Controlling pest populations entails subjecting them to very heavy selective pressures. It is obvious that a part of the population will escape the effects of these pressures thanks to polymorphism. Thus several selective pressures should be exerted on a population at a given time, and the combination of these pressures should be changed randomly in time. There is no escape to this state of affairs, even for control measures based on insecticide use!

CONCLUSION

In conclusion, this brief review of the behaviour and functioning of sense organs shows the complexity of the relations between Bruchids and their host plant and therefore the high level of adaptation of these pests to their food base. This is by no means an original conclusion. The mere fact that living beings exist undoubtedly proves that they are adapted. Hence, the problem is not to repeat this statement but to analyse the mechanisms of adaptation. This is the only way to design control methods which do not pose threats to our own existence both in the medium and long term.

REFERENCES

Arnoux, J., Labeyrie, V. & Maison, P., 1958. Influence de la température et de l'insolation sur la migration des Bruches du haricot (*Acanthoscelides obtectus* Say) dans la nature. C.R. Acad. Sci. 247: 2343–2345.

Autrum, H., 1977. Concept and method in sensory physiology. J. comp. Physiol. 120: 87–100.
Dethier, V.G., 1954. Evolution of feeding preferences in phytophagous insects. Evolution 8: 33–54.
Dethier, V.G., 1974. Sensory imput and the inconstant fly, pp. 21–34. In: 'Experimental analysis of insect behaviour' (Barton-Browne, ed.). Springer-Verlag, Heidelberg.
Fraenkel, G.S., 1959. The raison d'être of secondary plant substances. Science 129: 1466–1470.
Fraenkel, G.S., 1969. Evaluation of our thoughts on secondary plant substances. Entomol. Exp. & Appl. 12: 473–86.
Harborne, J.R., Boulter, D. & Turner, B.L., 1971. Chemotaxonomy of the *Leguminosae.* Academic Press, London & New York.
Kennedy, J.S., 1977. Olfactory responses to distant plants and other odor sources, pp. 67–91 and: Behaviourally discriminating assays of attractants and repellents, pp. 215–229. In: Chemical control of insect behavior: Theory and practice (H.H. Shorey & J.J. McKelvey, eds.). Academic Press, London & New York.
Labeyrie, V. & Maison, P., 1954. Sur les relations entre la ponte d'*Acanthoscelides obtectus* Say dans la nature et les stades phénologiques de *Phaseolus vulgaris* Lin. C.R. Acad., Sci. 238: 1920–1922.
Mayr, E. (1958) Behavior and systematics, pp. 241–262. In: Behavior and Evolution, A. Roe and G.G. Simpson, eds. Yale Univ. Press, New Haven.
Pajni, H.R. & Sood, S., 1975. Effect of peapollen feeding on maturation and copulation in the beetle *Bruchus pisorum* L. Indian J. of exp. Biol. 13, 202–203.
Perttunen, V. & Hayrinen, T., 1969. The effect of temperature and flight initiation and take-off rate in *Acanthoscelides obtectus* Say (Col. Bruchidae) Ann. Ent. Fenn., 35, 190–204.
Perttunen, V. & Hayrinen, T. (1970) Individual variations in the take-off activity of *Acanthoscelides obtectus* Say Ann. Ent. Fenn. 36, 107–110.
Pouzat, J., 1970. Rôle des organes sensoriels céphaliques dans l'ovogenèse et l'émission chez la Bruche du haricot *Acanthoscelides obtectus* Say. In: 'L'influence des stimuli externes sur la gamétogenèse des insectes'. Coll. Int. C.N.R.S., 189, 381–400.
Pouzat, J., 1974a. Comportement de la bruche du Haricot femelle (*Acanthoscelides obtectus* Say, Coleoptera: Bruchidae) soumise à différents stimulus olfactifs. C.R. Acad. Sci., Paris (D) 278: 2173–2176.
Pouzat, J., 1974b. Electro-antennogrammes de Bruche du haricot femelle (*Acanthoscelides obtectus* Say, Coleoptera: Bruchidae) soumise à différents stimulus olfactifs (dont la plante-hôte, le haricot et le mâle). C.R. Acad. Sci., Paris (D) 278, 2323–2326.
Pouzat, J., 1975. Analyse expérimentale du rôle de l'ovitube dans le comportement de ponte de la Bruche du haricot (*Acanthoscelides obtectus* Say, Coleoptera, Bruchidae). Behaviour 54 (3–4), 258–277.
Pouzat, J., 1976. Le comportement de ponte de la Bruche du Haricot en présence d'extrait de plante-hôte. Mise en évidence d'interactions gustatives et tactiles. C.R. Acad. Sci., 282, 1971–1974.
Pouzat, J., 1977. Effets des stimulations provenant de la plante-hôte, le haricot (*Phaseolus vulgaris* L) sur le comportement de ponte de la bruche du haricot (*Acanthoscelides obtectus* Say). Coll. Int. C.N.R.S., 265, 115–131.
Pouzat, J., 1978a. La régulation de l'activité reproductrice des femelles de Bruche du haricot (*Acanthoscelides obtectus* Say): Influence de quelques facteurs externes (plante hôte, activité locomotrice). Thèse d'Etat, Université F. Rabelais Tours.
Pouzat, J., 1978b. Host plant chemosensory influence on oogenesis in the bean weevil *Acanthoscelides obtectus,* (Coleoptera: Bruchidae) Ent. Exp. & Appl. 24: 401–408.
Slobodkin, L.B., 1968. Towards a predictive theory of evolution, pp. 187–205. In: Population, biology and evolution, (R.C. Lewontin, ed.). Syracuse Univ. Press, Syracuse, N.Y.
Visser, J.H., 1979. Olfaction in the Colorado Beetle. At the onset of host plant selection. Pudoc, Wageningen.
Wigglesworth, V.B., 1965. The principles of Insect Physiology. Methuen, London.
Wilson, E.O. & Bossert, W.H., 1963. Chemical communication among animals Rec. Prog. Hormone Res. 19, 673–716.

Zachariae, G., 1958. Das Verhalten des Speisebohnenkäfers *Acanthoscelides obtectus* Say (Coleoptera Bruchidae) im Freien in Norddeutschland. Zeit. ang. Entomol., 43, 345–365.

Author's address:
I.B.E.A.S. (ERA 328, CNRS)
Faculté des Sciences
Université de Tours
Parc de Grandmont
37200 Tours
France

RELATIONS OF *ACANTHOSCELIDES* WITH THEIR PLANT HOSTS

C.D. JOHNSON
(*Flagstaff, Ariz., U.S.A.*)

INTRODUCTION

Because it contains more species than any other genus of New World bruchids, I think a discussion of the history and size of *Acanthoscelides* is appropriate before I discuss its host relations.

The first revision of this New World genus was performed by Johnson (1970). I am now studying the ecosystematics of the genus in southern Mexico and Central America and it appears that it is likely to be split into several genera. At present, *Acanthoscelides* is estimated to contain about 60 named species in the United States, 58 in Mexico, 37 in Central America, and 120 in South America, the West Indies and the Galapagos. With overlap between regions, there are some 240 named forms of *Acanthoscelides* in the New World with more than 100 new species yet to be named. It is by far the largest genus of New World Bruchidae.

HOST PLANTS

According to Johnson (1970) 84% of bruchids feed in the seeds of the Leguminosae. Johnson (1981) listed 31 families in addition to the Leguminosae that are fed upon by bruchids. Recently, Kingsolver (1979) reported that an additional family, the Nymphaceae, has its seeds fed upon by the bruchid *Callosobruchus chinensis.*

A host plant of Bruchidae has long been considered by bruchid specialists to be plants that provide nourishment for the development of the larvae. Unfortunately, records of adults collected *on* plants have been published as 'hosts'. Thus, it is sometimes difficult to establish what constitutes a 'host' when reading published lists. The list published by Zacher (1952) should be used with extreme care because he did not indicate the source of his data. Johnson (1981) published the most accurate, up-to-date list of hosts but since he incorporated some records of Zacher and others whose host records are suspect, this list also should be used with caution.

The host list for *Acanthoscelides* (Johnson, 1981), however, I believe is very accurate with the possible exception of the many hosts (20 species in 11 genera) listed for *Acanthoscelides obtectus.* These hosts must be verified by researchers in economic entomology and probably those in the Old World (e.g., Southgate, 1978).

Species of *Acanthoscelides* are known to feed in seeds of nine families of plants. Most feed in seeds of the Leguminosae, the third largest family of plants after the Compositae and Orchidaceae, with some 17 000 species (Heywood, 1971). The three legume subfamilies are the Papilionoideae, Mimosoideae, and the Caesalpinioideae (Heywood, 1971). The other host families (Anacardiaceae, Cistaceae, Lythraceae, Malvaceae, Onagraceae, Rhamnaceae, Sterculiaceae, and Tiliaceae) have 160 to 1000 species in them (Munz and Keck, 1959).

Of the nine families, the Malvaceae, Sterculiaceae and Tiliaceae are in the same order (Malvales); the Lythraceae and Onagraceae are both in the order Myrtales, while the Anacardiaceae is in the order Rutales which is related to Myrtales. There are no other close relationships between these families (Takhtajan, 1969).

BRUCHIDS IN SEEDS OF PHASEOLUS

The apparent original and still principal host of the bean weevil, *A. obtectus*, is *Phaseolus vulgaris* and other species of *Phaseolus.* Three other species of *Acanthoscelides* are known to feed in species of *Phaseolus.* Of these, *A. obvelatus* and *A. argillaceus* (= *A. obreptus*) are very similar in their color and structure to *A. obtectus* and form a tightly-knit species group. The other species is *A. clandestinus* (= *A. multisignatus*) and is neither closely related to *A. obtectus* nor very common or widespread in seeds of species of *Phaseolus. Zabrotes subfasciatus*, a New World genus and species, is a common pest in commercial beans. Three species of the Old World genus *Callosobruchus* are also reported to feed in seeds of *Phaseolus.*

SPECIES OF ACANTHOSCELIDES THAT FEED IN VARIOUS HOSTS

When absolute numbers of the records of Johnson (1981) are listed under their host families and subfamilies and analyzed (Table 1), 85.5% of the species are shown to prefer the Leguminosae, with the remaining 14.5% in the eight non-legume families. Preferences for papilionoid legumes is indicated in that 52.1% of all host records are from this subfamily. The seeds of the Mimosoideae constitute 31.3% of all records. The Caesalpinioideae have about as many records (2.1%) as most of the non-legumes. Johnson (1981) indicated that seeds of this subfamily in general support fewer bruchids than do the other two subfamilies. The family Malvaceae has a greater number of species that feed in its seeds than the other non-legumes.

Most bruchids are host specific to genera of plants (Table 4; Johnson, 1981), but *Acanthoscelides compressicornis* is the only species in the genus known to feed in two subfamilies of legumes. None feeds in more than one family of plants.

Table 1. Absolute numbers and percentage of species of *Acanthoscelides* that feed in seeds of their various host families and subfamilies.

Family	Number of species	% Feeding in taxon
Leguminosae		
Papilionoideae	50	52.1
Mimosoideae	30*	31.3
Caesalpinioideae	2*	2.1
Subtotal (legumes)	82	85.5
Anacardiaceae	1	1.0
Cistaceae	3	3.1
Lythraceae	1	1.0
Malvaceae	4	4.2
Onagraceae	1	1.0
Rhamnaceae	1	1.0
Sterculiaceae	1	1.0
Tiliaceae	2	2.1
Subtotal (non-legumes)	14	14.5
Total	96	

**A. compressicornis* feeds in seeds of both subfamilies.

When absolute numbers of records and percentages are calculated only for *Acanthoscelides* that feed in the Leguminosae (Table 2), it is even more evident that species of *Acanthoscelides* prefer the seeds of papilionoid legumes (61%), with fewer in the Mimosoideae (36.6%) and hardly any in the Caesalpinioideae (2.4%).

When ratios are used, however, a different picture emerges. If the numbers of species of *Acanthoscelides* known to infest the three subfamilies are divided into the number of plant species the following results are obtained: Papilionoideae (ca. 12 000 spp.) 1 *Acanthoscelides*: 240 spp. papilionoid legume species; Mimosoideae (ca. 2500 spp.) 1 : 83; and Caesalpinioideae (ca. 2500 spp.) 1 : 1250. Thus, in terms of their availability, the mimosoid legumes are the preferred hosts of species of *Acanthoscelides.*

An 'exploitation number' (Table 3) gives an index of adaptive radiation of species of *Acanthoscelides* into the seeds of the genera and species of Leguminosae. This technique emphasizes the number of species of plants that are fed upon by *Acanthoscelides* in each genus. As with the other methods using percentages, the Papilionoideae are preferred (64.9%) more than the other two subfamilies (Mimosoideae: 31.7%; Caesalpinioideae: 3.4%), and *Astragalus* and *Mimosa* surface as the most exploited genera, with *Dalea*, *Phaseolus*, *Rhynchosia*, *Desmanthus* and *Leucaena* also having many of their species fed upon by species of *Acanthoscelides.*

HOST SPECIFICITY

Several authors (e.g., Janzen, 1975a, 1977, 1978; Johnson & Slobodchikoff, 1979) have discussed the host specificity of the Bruchidae. It has been

Table 2. Absolute numbers and percentage of species of *Acanthoscelides* that feed in seeds of subfamilies of the Leguminosae.

Subfamily	Number of species	% Feeding in taxon
Papilionoideae	50	61.0
Mimosoideae	30*	36.6
Caesalpinioideae	2*	2.4
Total	82	

**A. compressicornis* feed in seeds of both subfamilies

evident that most bruchids have one or few host plants and that most hosts are fed upon by one or a few bruchids. For example Johnson & Slobodchikoff (1979) found that 82.5% of the bruchid species that attack species of the legume genus *Cassia* had only one to three hosts whereas 16.5% of the bruchids had five to ten hosts. They also found that 90.5% of the species of *Cassia* support three or fewer species of bruchids. The remaining species of *Cassia* support from five to seven species of bruchids.

In this paper, similar results are presented (Table 4). Almost 50% of the species of *Acanthoscelides* are known to have only one host and almost 70% have one to two hosts (Table 4). The larger genus *Acanthoscelides*, however, has six species with more than one host and two with more than 20.

Similar results were evident when the hosts were examined. Almost 60% of the host genera support only one species of *Acanthoscelides* and 74% support one to two species of these bruchids (Table 4). Almost 90% of the host genera of *Acanthoscelides* support from one to four bruchid species while only *Mimosa* has as many as 12 species in its seeds.

LIFE HISTORIES OF ACANTHOSCELIDES

The data available from the literature on *Acanthoscelides* life histories are presented in Table 5 and indicate that most species of *Acanthoscelides* (70%) oviposit on the pod surface while only 29% oviposit directly on seeds. Some, however, such as *chiricahuae* and *obtectus* may oviposit on both seeds and pods. An overwhelming 75% feed and pupate inside a single seed. Several species (*baboquivari*, *collusus*, *prosopoides*, *submuticus*), however, feed in the seeds of plants that have only one seed per fruit. It has been suggested by Center & Johnson (1974) that those bruchids whose larvae feed on several seeds in the course of their development are adapting to a coevolved trait, i.e., the plants are reducing their seed size in an attempt to exclude the bruchids.

The number of generations per year is of importance to stored-product entomologists because these species may become economic problems. Only five of the species in Table 5 are known to continue to breed in the laboratory. Probably more would breed continuously if optimum conditions and abundant food were provided for them.

Table 3. Exploitation or adaptive radiation of *Acanthoscelides* spp. in genera and species of Leguminosae. This index was obtained by counting the numbers of bruchid species – plant species combinations. Thus *A. aureolus* feeds in seeds of 19 species of *Astragalus* and was given a total of 19 points in *Astragalus. A. aureolus* also feeds in two species of *Oxytropis* so this genus was scored as two. Other 'exploitation numbers' were calculated in the same way.

	'Exploitation number'	% of subfamily fed upon by *Acanthoscelides* (by genus)	% of family fed upon by *Acanthoscelides* (by genus)
Genera of Papilionoideae			
Aeschynomene	5	2.9	1.9
Amorpha	6	3.4	2.2
Astragalus	53	30.5	19.8
Baptisia	1	0.6	0.37
Cajanus	2	1.2	0.74
Calopogonium	3	1.7	1.1
Cicer	1	0.6	0.37
Colutea	2	1.2	0.74
Dalea	15	8.6	5.6
Desmodium	9	5.2	3.4
Eriosema	1	0.6	0.37
Errazurizia	1	0.6	0.37
Eysenhardtia	1	0.6	0.37
Galactia	2	1.2	0.74
Glycine	1	0.6	0.37
Glycyrrhiza	2	1.2	0.74
Hedysarum	1	0.6	0.37
Indigofera	6	3.4	2.2
Lablab	1	0.6	0.37
Lathyrus	1	0.6	0.37
Lens	1	0.6	0.37
Lespedeza	4	2.3	1.5
Lotus	5	2.9	1.9
Mucuna	1	0.6	0.37
Nissolia	2	1.2	0.74
Oxytropis	2	1.2	0.74
Parryella	1	0.6	0.37
Phaseolus	10	5.8	3.7
Pisum	1	0.6	0.37
Psorothamnus	3	1.7	1.1
Rhynchosia	12	6.9	4.5
Sesbania	3	1.7	1.1
Tephrosia	6	3.4	2.2
Trifolium	3	1.7	1.1
Vicia	1	0.6	0.37
Vigna	5	2.9	1.9
Subtotal	174		64.9
Genera of Mimosoideae			
Acacia	4	4.7	1.5
Desmanthus	19	22.4	7.1
Leucaena	11	12.9	4.1
Mimosa	39	45.9	14.6
Parkia	2	2.4	0.74
Piptadenia	1	1.2	0.37

Table 3 (*continued*)

	'Exploitation number'	% of subfamily fed upon by *Acanthoscelides* (by genus)	% of family fed upon by *Acanthoscelides* (by genus)
Prosopis	4	4.7	1.5
Schrankia	5	5.9	1.9
Subtotal	85		31.7
Genera of Caesalpinoideae			
Cassia	6	66.7	2.2
Hoffmannseggia	3	33.3	1.1
Subtotal	9		3.4
Total	268		

Table 4. Number of species of *Acanthoscelides* and their larval hosts, and number of host genera with *Acanthoscelides* predators. Column 1 is the number of *Acanthoscelides* species that attack a certain number of hosts (column 2). Column 3 is the percentage of total species of *Acanthoscelides* attacking that many host species. For example, there are 47 species of *Acanthoscelides* that each attack only a single host, or 49.5% of the total *Acanthoscelides* species attack only a single host species. Column 4 is the number of host genera that support a certain number of *Acanthoscelides* species (Column 5), and column 6 is the percentage of total host genera supporting that many *Acanthoscelides* species. For example, there are 34 plant genera that each support only a single species of *Acanthoscelides*, or 58.6% of the host genera are attacked by only a single *Acanthoscelides* species.

(1)	(2)	(3)	(4)	(5)	(6)
47	1	49.5	34	1	58.6
18	2	18.9	9	2	15.5
7	3	7.4	5	3	8.6
7	4	7.4	4	4	6.9
4	5	4.2	2	5	3.4
2	6	2.1	2	6	3.4
3	7	3.2	1	7	1.7
0	8	0	–	–	–
1	9	1.1	1	12	1.7
0	10	0			
2	11	2.1			
1	12	1.1			
–	–	–			
1	16	1.1			
–	–	–			
1	20	1.1			
–	–	–			
1	24	1.1			

Table 5. Traits of some *Acanthoscelides.* (compiled from Janzen, 1975b; Johnson, 1970, 1974, 1977a, b; Center & Johnson, 1974; Forister & Johnson, 1971; Forister, 1970; Johnson & Kingsolver, 1971; Pfaffenberger & Johnson, 1976).

	Oviposition		Feeding		Pupation		Number of Generations/year
Acanthoscelides	on pod	on seed	one to many seeds	one seed	outside one seed	inside one seed	
alboscutellatus	?	?	X		X		?
aureolus		X		X		X	2 +
baboquivari	X			X		X	many
chiricahuae	X	X		X		X	many
compressicornis	X		X		X		1–2
fraterculus	X		X		X		1–2
guazumae		X		X		X	1
kingsolveri	X			X		X	many
lobatus	X		X		X		1?
mixtus	X			X		X	1?
mundulus	X			X		X	many
obtectus	X	X		X		X	many
pallidipennis	X			X		X	1
pauperculus		X		X		X	1?
prosopoides	X			X		X	3
submuticus	X			X		X	1
Total	12	5	4	12	4	12	
%	70.6	29.4	25	75	25	75	

PEST POTENTIAL OF ACANTHOSCELIDES

The species other than *A. obtectus* that have the greatest numbers of hosts (*A. aureolus*, *A. chiricahuae*, *A. compressicornis*, *A. distinguendus*, *A. fraterculus*, *A. mexicanus*, *A. pullus*) could become pests, but their host seeds are not of any particular economic importance at this time. Most are fairly specific to genera of legumes so large-scale host transfers seem unlikely.

Species that feed in seeds of species of *Phaseolus* seem to be the most likely candidates to become pests. In addition to *A. obtectus*, *A. obvelatus*, *A. argillaceus*, and *A. clandestinus* feed in seeds of species of *Phaseolus.* If these species were introduced into a foreign, more suitable habitat along with *Phaseolus*, there is the potential for them to become pests.

Acanthoscelides zeteki has been reared from *Rhynchosia minima* and *Flemingia strobilifera* (Johnson, unpublished) from a latitudinal range from Mazatlan, Mexico, to Panama, and it also attacks the seeds of the cultivated *Cajanus cajan* (Kingsolver, 1969). Southgate (1979) has reported that this bruchid has become a widespread pest of stored *C. cajan* in the Caribbean in recent years. *A. flavescens*, a species closely related to *A. zeteki*, also feeds in species of *Rhynchosia* in Mexico and Central America.

Other species which have wide distributions in Latin America and could become pests because of their continuous breeding habits are *A. rufovittatus*, which feeds in *Tephrosia cinerea* (Johnson, unpublished), and *A. puelus*, which is widespread in the seeds of *Calopogonium mucunoides.* Their combined distributions are from Arizona to Panama.

There is potential for host transfer by *A. compressicornis* because it has at least 11 hosts in two subfamilies, but as yet has not been recorded from economically important seeds.

Perhaps the species of *Acanthoscelides* with the greatest potential to become pests are those that attack the seeds of clovers (*Trifolium*). The species that are known to infest clover seeds are *A. pauperculus*, *A. pusillimus* and *A. pyrrhomelas*. A host transfer to introduced clover seeds appears to have already been made as the latter two species are reported from the introduced Eurasian red clover, *Trifolium pratense*. Because *A. pauperculus* is a very abundant, morphologically variable and widespread species in the western United States, it seems very probable that it will eventually be found to feed in clover seeds of economic importance.

Those species of *Acanthoscelides* that feed in seeds of the Malvaceae have the potential to invade seeds of such economically important plants as cotton (*Gossypium*), okra (*Abelmoschus*), and hibiscus (*Hibiscus*). All three plant genera have been reported to support bruchids in genera other than *Acanthoscelides*. The record from *Gossypium* must be verified, however.

The important legume genus *Leucaena* is known to be a nutritious forage and organic fertilizer as well as having tree types that produce timber, firewood, pulp and paper. *Leucaena* species are also valuable for revegetation, providing firebreaks, shade, and city beautification (National Academy of Sciences, 1977). The green pods are also used as food for humans in parts of Mexico and Central America (personal observation). Because several species of *Acanthoscelides* are known to feed in seeds of species of *Leucaena*, these bruchids may become economically important as the use of *Leucaena* progresses.

ACKNOWLEDGEMENTS

I am grateful to the U.S. Department of Agriculture and the National Science Foundation for several grants which provided funds for gathering many of the host records.

REFERENCES

Center, T.D. & Johnson, C.D., 1974. Coevolution of some seed beetles (Coleoptera: Bruchidae) and their hosts. Ecology 55: 1096–1103.

Forister, G.W., 1970. Bionomics and ecology of 11 species of Bruchidae (Coleoptera). M.S. Thesis, Northern Arizona University. 93 pp.

Forister, G.W. & Johnson, C.D., 1971. Behaviour and ecology of *Acanthoscelides prosopoides* (Coleoptera: Bruchidae). Pan-Pac. Entomol. 47: 224–234.

Heywood, V.H., 1971. The Leguminosae-A systematic purview. In: Chemotaxonomy of the Leguminosae. (J.B. Harborne, D. Boulter & B.L. Turner, eds.). Academic Press, London & New York. 612 pp.

Janzen, D.H., 1975a. Interactions of seeds and their insect predators/ parasitoids in a tropical deciduous forest. In: Evolutionary Strategies of Parasitic Insects and Mites (P.W. Price, ed.). Plenum, New York.

Janzen, D.H., 1975b. Intra- and interhabitat variations in *Guazuma ulmifolia* (Sterculiaceae) seed predation by *Amblycerus cistelinus* (Bruchidae) in Costa Rica. Ecology 56: 1009–1013.

Janzen, D.H., 1977. The interaction of seed predators and seed chemistry, p. 415–428. In: Colloques Internationaux du C.N.R.S., Paris (V. Labeyrie, ed.). 493 pp.
Janzen, D.H., 1978. The ecology and evolutionary biology of seed chemistry as relates to seed predation, pp. 163–206. In: Biochemical Aspects of Plant and Animal Coevolution (J.B. Harborne, ed.). Academic Press, London.
Johnson, C.D., 1970. Biosystematics of the Arizona, California, and Oregon species of the seed beetle genus *Acanthoscelides* Schilsky (Coleoptera: Bruchidae). Univ. Calif. Publ. Entomol. 59: 1–116.
Johnson, C.D., 1974. Ecology of two *Acanthoscelides* from *Indigofera*, with a description of a new species (Coleoptera: Bruchidae). J. Kans. Entomol. Soc. 47: 268–278.
Johnson, C.D. 1977a. Ecology and behaviour of *Acanthoscelides mundulus* in seeds of *Nissolia schottii* (Coleoptera: Bruchidae; Leguminosae). Pan-Pac. Entomol. 53: 161–167.
Johnson, C.D., 1977b. Notes on the host plants and distribution of *Acanthoscelides pauperculus* (Le Conte)(Coleoptera: Bruchidae). Pan-Pac. Entomol. 53: 303–304.
Johnson, C.D. 1981. Seed beetle host specificity and the systematics of the Leguminosae. In: Advances in Legume Systematics (R.M. Polhill and P.H. Raven eds.)(in press).
Johnson, C.D., & Slobodchikoff, C.N., 1979. Coevolution of *Cassia* (Leguminosae) and its seed beetle predators (Bruchidae). Environ. Entomol. 8: 1059–1064.
Kingsolver, J.M. 1979. A new host record for *Callosobruchus chinensis* (L.)(Coleoptera: Bruchidae). Coleopterists Bull. 33: 438.
Munz, P.A. & Keck, D.D., 1959. A California Flora. University of California Press, Berkeley. 1681 pp.
National Academy of Sciences, 1977. *Leucaena*: Promising Forage and Tree Crop for the Tropics. 118 p.
Pfaffenberger, G.S. & Johnson, C.D., 1976. Biosystematics of the first-stage larvae of some North American Bruchidae (Coleoptera). U.S. Dept. Agric. Tech. Bull. 1537: 101 p.
Southgate, B.J., 1978. The importance of the Bruchidae as pests of grain legumes, their distribution and control. In: Pests of Grain Legumes: Ecology and Control (S.R. Singh, H.F. van Emden & T.A. Taylor eds.). Academic Press, London. 454 p.
Southgate, B.J., 1979. Biology of the Bruchidae. Ann. Rev. Entomol. 24: 449–473.
Takhtajan, A., 1969. Flowering Plants Origin and Dispersal. Smithsonian Institution Press, Washington, D.C. 310 p.
Zacher, F., 1952. Die Nährpflanzen der Samenkäfer. Z. Angew. Entomol. 33 (3): 460–480.

Author's address:
Dept of Biological Sciences
Northern Arizona University
Flagstaff
Arizona 86011
U.S.A.

FIRST DATA ON BRUCHIDAE WHICH ATTACK THE PODS OF LEGUMES IN UPPER VOLTA OF WHICH EIGHT SPECIES ARE MAN CONSUMED

C. VARAIGNE-LABEYRIE & V. LABEYRIE
(Ouagadougou, Haute Volta/Tours, France)

INTRODUCTION

The Saria area, 12° north, in the northern part of the Guinea-Sudanese land, at an altitude of 300 m, has a mean annual 800–850 mm rainfall. The precipitation concentrated from late May till October and above all the importance of diurnal evapotranspiration throughout the dry season (with a temperature always above 35°C in the hottest hours) result in a vegetation with a very marked seasonal rhythm.

This agricultural area of the Mossi Land being densely populated (over 50 inhabitants per square kilometre) has resulted in a total utilization of rural space, and in an essentially anthropogenic vegetation which has been frequently altered, with mega and mesophanerophytes becoming rare as a consequence of the systematic utilization of biomass as fuel. In that vegetation, legumes are very important out of 29 legume species whose abundance allowed sampling, 6 are mesophanerophytes, 3 of which at least were introduced; 13 microphanerophytes contribute to give the vegetation its characteristic aspect; 2 are nanophanerophytes lastly, 8 species are therophytes.

While that area of ancient agrarian civilization has been subjected to a repeated deterioration of its natural plant cover, various species have been brought into, following the great routes and lines of exchange of the precolonial period, and there have been many more cosmopolitan introductions at the time of colonization.

The infestation of legume seeds by bruchids can give indications about the chemical resemblances of seeds of various species, and the integration of the various legumes in the local biocenoses.

In effect, indigenous bruchid species rarely colonize introduced legumes. For example in Japan, *Vicia Faba* L. pods are not attacked by any local weevil, and the first damage occurred only about 1925 when *Bruchus rufimanus* Boh was accidentally introduced (Kamito, 1933). Thus, even the attack of the ripe pods of *Arachys hypogea* L., after it was introduced into Africa, by *Caryedon serratus* 01., must not be delusive, because they are not attacked in situ, but after digging out, when they are exposed for a long time to dry.

Series Entomologica, Vol. 19: ed. by V. Labeyrie

CONDITIONS OF STUDY

Over 3000 pods were collected, mostly in the Saria area, or a few of them in Ouagadougou, on a variable, but limited number of plants belonging to 29 spontaneous or cultivated legume species.

The number of collected pods per species varied from 8 for *Piliostigma thonningii* to 328 for *Cajanus cajan.* As the pods of 6 species (*Acacia sieberiana*, *Cassia sieberiana*, *Delonix regia*, *Parkia biglobosa*, *Piliostigma reticulatum* et *P. thonningii*), were longer than the available polyethylene bags, they were kept in muslin cases with 1 mm meshes. So a few insects may have escaped.

As there were no available data on the phenology and the persistence of the fruits of these legumes, the sampling of each species could not be planned, *as in any study of that kind.*

So the 29 legume species were collected from one to six times each, according to the persistence of their fruits, between September 1976 and August 1977, in 70 pickings.

Number of sampled species:	9	5	6	4	1	2
Number of pickings :	1	2	3	4	5	6

However limited this type of work* may be, it provided information on the relationships between legumes and bruchids in one part of the world, where only Prevett's observations in Nigeria were available (1966a, 1967).

The data on plants are related to the dates of flowering, ripening and dehiscence of fruits. The duration of the ripening period of fruits, how long they stay on the tree; the way the seeds are released.

The data on bruchids concern the distribution of attacks per pod and per seed, the sites of larval development and pupation, the date of emergence of adults, the number of generations and the possibilities of re-infestation in situ, and the entomophagous insects obtained.

OBSERVATION OF THE RIPENING OF LEGUMES

Out of 29 species studied, 8 are man-consumed. For seven of them, the seeds are: *Acacia seyal* Del., *Arachis hypogea* L., *Cajanus cajan* L., *Parkia biglobosa* Benth., *Piliostigma thonningii* M., *Vigna unguiculata* L., *Voandzeia subterranea* L.; for *Tamarindus indica* L., it is the pulp.

The observations were made on 8 caesalpiniaceae, 13 mimosaceae, and 8 papilionaceae.

Mesophanerophytes: mimosaceae (*Acacia nilotica* L., *Albizia lebbeck* L.,

*This work was carried out by C. Varaigne-Labeyrie when she was a professor at the 'Institut Superieur Polytechnique' in Ouagadougou and was presented as a 'Diplôme d'Etudes Approfondies' at the F. Rabelais University, Tours, in 1978.

The information on legumes was added to by O. Bougnounou and S. Ouattara's observations, from the 'Centre Voltaique de la Recherche Scientifique' and J.F. Duranton from the G.E.R.D.A.T. The determination of bruchids were made by K. Decelle, Director of the Entomology Laboratory of the Royal Museum of Central Africa in Tervuren, Belgium.

Parkia biglobosa Benth.); Caesalpiniaceae (*Delonix regia* Raf., *Tamarindus indica* L.); papilionaceae (*Pterocarpus erinaceus* Poir.)

Microphanerophytes: mimosaceae (*Acacia dudgeoni* Graib., *A. gourmaensis* A. Ch., *A. macrostachya* Reichenb., *A. pennata* Graib., *A. seyal* Del., *A. sieberiana* D.C., *Dicrostachys glomerata* Chiov., *Entada africana* G., *Leucaena glauca* L.); Caesalpiniaceae (*Cassia sieberiana* D.C., *Piliostigma thonningii* Milne, *P. reticulatum* Hochst, *Parkinsonia aculeata* L.)

Nanophanerophytes: Mimosaceae (*Mimosa pigra* L.) papilionaceae (*Cajanus cajan* L.)

Therophytes: Caesalpiniaceae (*Cassia occidentalis* L. and *C. tora* L.) Papilionaceae (*Arachys hypogea* L., *Crotalaria retusa* L., *Sesbania pachycarpa* D.C., *S. sesban* L., *Vigna unguiculata* L., *Voandzeia subterranea* L.)

All of them form pods, except *Pterocarpus erinaceus*, which bears winged seeds.

The pods of *Arachis hypogea* and *Voandzeia subterranea* are hypogean.

Out of 26 species with epigean pods, 19 remain indehiscent on the plants. The pods of 2 of these species (*Cassia sieberiana* and *Mimosa pigra*) divide into articles. The pods of *Cassia tora* and *Albizia lebbeck* only open when the rainy season comes.

So the pods of the other species are closed when they fall to the ground: those of *Acacia dudgeoni* (most of them), *A. macrostachya*, *A. pennata* and *Leucaena glauca* (most of them) open on the ground; the pods of *Acacia nilotica*, *A. sieberiana*, *Cajanus cajan*, *Delonix regia*, *Dichrostachys glomerata*, *Parkinsonia aculeata*, *Parkia biglobosa*, *Piliostigma reticulatum*, *P. thonningii*, *Tamarindus indica* and *Vigna unguiculata* progressively decay on the ground.

Under such conditions, only the pods of 7 species (*Acacia gourmaensis*, *A. seyal*, *Cassia occidentalis*, *Crotalaria retusa*, *Entada africana*, *Sesbania pachycarpa*, *S. sesban*) open on trees during the dry season; but the seeds of *A. seyal* remain attached to the ventral suture of the open pod. Therefore dehiscence can be a way of dispersal of the seeds of only 6 species.

The frequency of dehiscent therophytes of 0.5 versus 0.1 in the phanerophytes, and the winged seeds of *Entada africana* which get detached one after the other are not immediately released by dehiscence.

The 26 species with epigean fruits which stay closed on the plants (Table 1) can theoretically be attacked by the bruchids developing in ripe seeds, second group for Loukaniovitch (1957).

Acacia gourmaensis does not appear here, as the evolution of its pods could not be studied because it could not be found in the sites under study, and it could be collected only outside the studied areas.

In December, 18 species bear ripe fruits; they progressively decrease in number till the rainy season comes. Then dehiscence has already caused the seeds of *Crotalaria retusa*, *Sesbania pachycarpa* and *S. sesban* to fall to the ground, and harvesting has removed the pods of *Vigna unguiculata*.

3 species present a rather surprising period of fructification: *Pterocarpus erinaceus* and *Acacia seyal* ripen at the end of the dry season, respectively in April and from March till May; the ripening of *Parkia biglobosa* pods begins just before the first rainfalls and then goes on over most of that

Table 1. Presence of epigean fruits on legumes in Upper Volta. (*diagonal lines*: rainy season; *full lines*: most fruits on the tree; *dotted lines*: only a few fruits left; *crosses*: dates of pickings.).

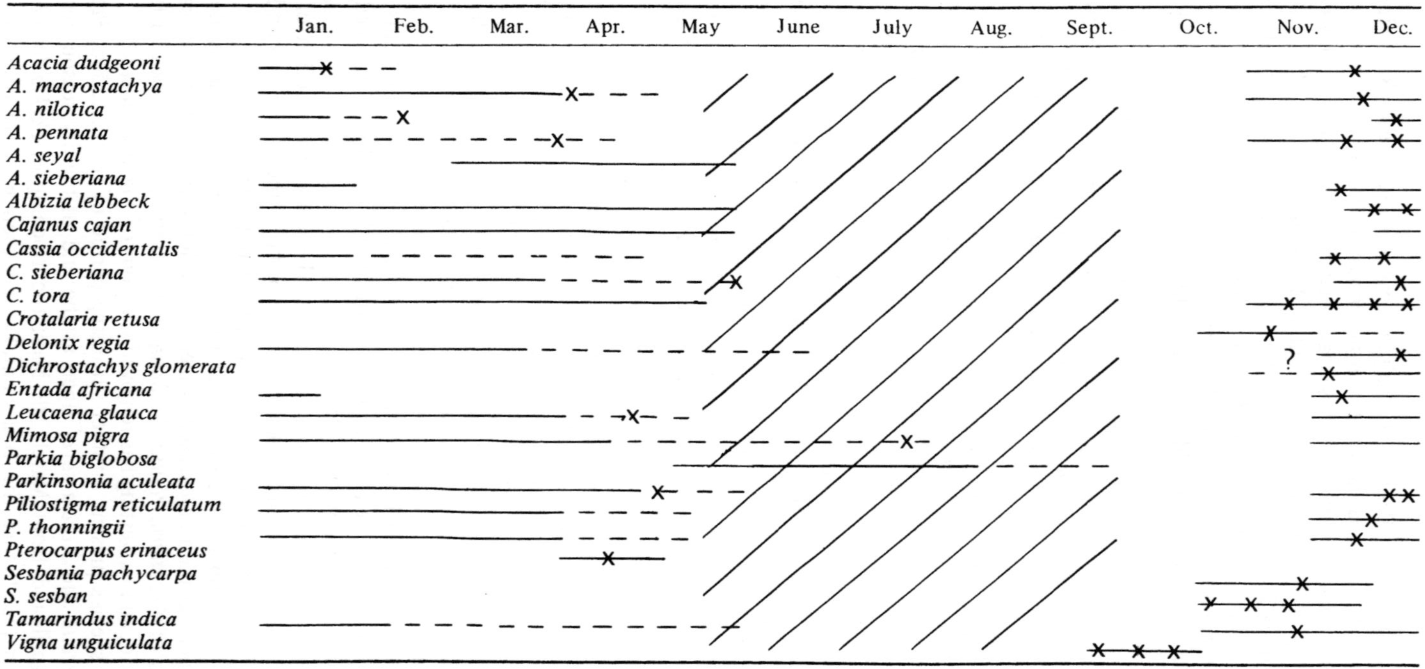

season. Could not this unseasonable cycle in comparison with all the other legumes, imply a more equatorial origin?

OBSERVATIONS ON THE ATTACKS OF BRUCHIDS

Table 2 shows all the data obtained from the collected pods. The number of adults obtained does not correspond to the number of emergence holes in the seeds; there is a deficit of 418, i.e. about 25%. It originates in 10 of the 19 legumes attacked. The sampling, which was sometimes started too late could not take into account the adults that flew away before the picking. It must account for all, or part of the deficit in *Acacia dudgeoni*, *A. macrostachya*, *A. seyal*, *Albizia lebbeck*, *Cassia sieberiana* and *Dichrostachys glomerata.*

Table 2. Data obtained from the collected pods.

Number of legume species	Number of pods examined	Number of seeds examined	Number of species attacked	Insect holes observed
29	3050	28–30 000	19	1635

Insects obtained	Number of bruchid species	Overall number of bruchids obtained	Bruchid species attacked by parasites	Number of hymenopterae obtained
1217	16	1197	4 to 7	20

Besides, the size of the meshes of the muslin bags in which the species with bigger pods were kept may have enabled some bruchids to escape from *Albizia lebbeck*, *Piliostigma reticulatum*, *P. thonningii* and *Tamarindus indica*. Lastly, some larvae of *Careydon interstinctus* may have escaped by boring the indehiscent pods of *Acacia nilotica* and *A. sieberiana* and the polyethylene bag.

Thus some bruchid species may have escaped that counting likewise, early emergences and escapes may have reduced the number and diversity of entomophagous insects.

Anyway, out of 16 species collected, 7 are new for the world fauna.

The distribution of the three families is given in Table 3: mimosaceae, (M), Caesalpiniaceae (C), and Papilionaceae (P) with respect to the number of bruchid species obtained.

Table 3.

Unattacked legumes	Legumes attacked by 1 bruchid species	Legumes attacked by 2 bruchid species	Legumes attacked by 3 bruchid species
10 = (4 M, 2 C, 4 P)	15 = (5 M, 6 C, 4 P)	2 = (2 M)	2 = (2 M)

The *Arachys hypogea* L. pods that were examined were taken from the plants, as the aim was the study of legume infestation in nature.

The mean rate of species which attack legumes is at least 25/29 = 0.86. As it was calculated from the cumulative number of bruchid species found in the pods, it is higher than the number of species, since some of them attack several legumes.

The rate of bruchid species is 0.5 for Papilionaceae, 0.75 for Caesalpinaceae, and 1.15 for Mimosaceae.

Table 4 sums up the data on the importance of attacks: number of bruchid species obtained per legume, number of exit holes to number of seeds ratio.

Prevett (1967), has also obtained 3 bruchid species from *A. sieberiana* but he did not collect *A. seyal.* In Nigeria, the legumes attacked by 2 species are *A. nilotica* and *Cassia sieberiana*, and not *A. macrostachya* and *Albizia lebbeck.*

The analysis of the attack rate number of holes of insect emergence to number of seeds ratio, does not give an absolutely precise indication for all legumes.

In effect, *Acacia macrostachya*, *Cassia occidentalis*, *C. sieberiana*, *C. tora*, *Piliostigma reticulatum*, *P. thonningii* and *Tamarindus indica* are attacked by polyvoltine *Caryedon*: so the exit holes can be those of insects from different generations having re-infested the same pods as a consequence of their being confined. Thus the attack rates can be overestimated. Inversely picking the pods of 19 species prior to their dehiscence an the fall of seeds can reduce the attacks.

Anyway the rate of attack must give a fairly accurate indication, and undoubtedly a much more reliable one than the number of species infesting the pods.

The 5 Mimosaceae are most attacked species, on the other hand no papilionaceae is to be found among the 8 most attacked species. The rates of attack of the 3 *Cassia* are quite similar, so are those of the 2 *Sesbania* and the 2 *Piliostigma.*

There is no connection between the rate of attack and the number of bruchid species contributing to these attacks; thus *Acacia seyal* which is attacked by 3 different species is 10 times less attacked than *A. pennata* is by one species.

Therefore the rate of attack, is not determined by the barrier of protective chemical substances which determines the number of bruchid species that infest a legume.

There is *no relationship between the size of seeds and the rate of attack*, contrary to what had been inferred by Janzen (1969) from his records in Central America. Lastly, though the pods which contain many seeds often show a rate of attack below 0.1, there is *no relationship between the number of seeds per pod and the level of attacks.* Thus *Cassia sieberiana* and *Piliostigma reticulatum* are both to be found in Saria, with 50 to 100 seeds per pod; but their rate of attack, which is respectively 0.004 and 0.09 is twentifold different.

On the other hand there is a much more conspicuous relationship between the intensity of attacks and the oldness of introduction of the legume into the Mossi area. Thus, out of 10 unattacked species (which are not recorded by Prevett in his list of legumes attacked in North Nigeria, 1967) 8 were probably introduced. They are: *Arachis hypogea* L., *Cajanus cajan* L. (which is being experimentally cultivated in Saria), *Crotalaria retusa* L. (which is cosmopolitan according to Zeven & Zhukovsky (1975) and used for its pigments in East Africa), *Delonix regia* Raf. (originating from Madagascar) *Leucaena glauca* Benth. (coming from Tropical America) *Mimosa pigra* (attacked by 2 *Acanthoscelides* species in Costa Rica (Janzen, 1977) whose American origin can be suspected) *Parkinsonia aculeata* (originating from Tropical America (Kerharo & Adam, 1971) and attacked in Costa Rica by 2 *Mimosestes*).

Parkia biglobosa Benth (The Mossis'nere) sets a much more difficult problem as it has been remarkably integrated into the national culture – it is consumed in the form of raw or fermented seeds (soumbala) and food pulp – it is very abundant, but on the other hand its completely dephased reproductive cycle might imply a more southern origin.

Cassia occidentalis and *C. tora* which are slightly attacked in Upper Volta are heavily attacked in Nigeria by the same *Caryedon* (Loukianovitch, 1957); Zeven & Zhutovsky (1975) describe them in their chapter devoted to the Indochinese–Indonesian centre of diversification.

Voandzeia subterranea, owing to it hypogean fructification is only attacked by some *Callosobruchus* which lay their eggs on the pods near the surface of the soil.

The six more attacked species are all considered as aboriginal ones.

The importance of attacks, though it is exceptionally high, apparently is detrimental neither to *Acacia macrostachya* (624 bruchids for 395 seeds) nor to *A. pennata* (131 bruchids for 262 seeds), which are very abundant species. Their medium-size seeds germinate even though they have been consumed by a weevil.

OBSERVATIONS ON BRUCHIDAE HARMFUL TO LEGUMES

The 16 collected species belong to 4 genera (Table 5). The hosts of a bruchid species belong to the same family and often to the same genus (Prevett, 1969). In the same way for Janzen (1977) in Costa Rica, 10 weevil species with several hosts attack the same genus, 12 the same family, to 2 attacking different families. For Prevett (1966b) in North Nigeria, 7 utilize one genus; 8 one family to 2 using different families.

The number of weevils obtained being conservatively estimated at about 1370, 890, i.e. two thirds, belong to species reccrded for the first time. Among those, *Bruchidius tougouriensis*, which was obtained from 3 different hosts, alone represents one third of the weevils collected during this study. It shows how limited our present knowledge is.

Out of 5 *Caryedon* species, 3 exclusively attack Mimosaceae. Decelle (1979) records *C. mauritanicus* in west Africa on *Acacia gourmaensis*, *A. dudgeoni*, *A. pennata* and *A. Seyal*. But in Upper Volta it was obtained only from *A. macrostachya*, though the seeds of these other legumes were also picked in large quantities. Prevett (1966b) records 2 other host caesalpiniaceae for *C. serratus*: *Cassia arerch* Del., and *Bauhinia rufescens* Lam., and 4 other

Table 4. Importance of attacks: number of bruchid species obtained by legume, number of exit holes to number of seeds ratio.

Un-attacked legumes 10		Legumes attacked by one bruchid species 15			Legumes attacked by 2 bruchid species 2			Legumes attacked by 3 bruchid species 2		
Mimosaceae:	Entada africana	Mimosaceaea	Acacia dudgeoni	xx	Mimosaceae	Acacia macrostachya	xxxxx	Mimosaceae	*Acacia seyal*	xx
4	Leucaena glauca	5	A. gourmaensis	xxx	2			2		
	Mimosa pigra		A. nilotica	xx		Albizia lebbeck	xx		A. sieberiana	xxx
	Parkia biglobosa		A. pennata	xxxx						
			Dickhrostachys glomerata	xxx						
Cesalpinaceae	Delonix regia	Cesalpinaceae	Cassia occidentalis	x						
2	Parkinsonia aculaeata	6	C. sieberiana	x						
			C. tora	x						
			Piliostigma reticulatum	xx						
			P. thonningii	xx						
			Tamarindus indica	xxx						
Paplionaceae	*Arachys hypogea*	Papilionaceae	Sesbania pachycarpa	xx						
4	*Cajanus cajan*	4	S. sesban	xx						
	Crotalaria retusa		*Vigna unguiculata*	xx						
	Pterocarpus erinaceus		*Voandzeia subterranea*	x						

(boxed) species whose number of bruchid's exit holes is at least equal to half the number of seeds.
(italic) man-consumed species

x number of exit holes < 0.01 number of seeds
xx number of exit holes from 0.01 to 0.1 of the number of seeds
xxx number of exit holes from 0.1 to 0.49 of the number of seeds
xxxx number of exit holes from 0.49 to the number of seeds
xxxxx number of exit holes higher than the number of seeds.

hosts for *C. pallidus*: 3 caesalpiniaceae (*Cassia italica* Mill., *C. mimosoides* L. and *C. singuena* Del. and one mimosaceae (*Acacia ataxacantha* DC).

Out of 9 *Bruchidius*, 8 have attacked mimosaceae; only *Bruchidius chloroticus* has attacked papilionaceae (*Sesbania pachycarpa* and *S. sesban*). On the other hand, out of the 11 species collected by Prevett (1966b) in Nigeria, 6 attack papilionaceae and 5 attack mimosaceae.

16 adults of 7 species were caught in spring (Table 6), 11 belong to species which were not obtained from parasitized pods. Conversely, 13 species having parasitized the pods were not caught, though they yielded 950 adults. For instance, neither *Caryedon mauritanicus*, nor *Bruchidius tougouriensis*, though it is very abundant, were caught on *A. macrostachya* whose seeds are parasitized by them. On the other hand, other bruchids were going about on these pods.

Caryedon pallidus and *Callosobruchus maculatus* are active in spring before and at the beginning of the rainy season. More surprising was the catching of *Bruchidius dichrostachydis*, apparently a univolting species whose adults yet must have emerged 6 months before.

9 species have a solitary development: *Bruchidius chloroticus*, *B. dichrostachydis*, *B. luteopigus*, *B. pennatae*, *B. silaceus*, *B. submaculatus*, *Caryedon interstinctus*, *C. pallidus* and *Tuberculobruchus natalensis.*

It is impossible to know if *Bruchidius sieberianae*, *B. voltaicus* and *Caryedon sahelicus* are definitely solitary because out of a hundred or so attacked *Acacia seyal* seeds, a few ones presented 2 bruchids exit holes. On the other hand the presence of a large number of *Bruchidius tougouriensis* larvae in the *Acacia macrostachya* seeds could have allowed the solitary development of *Caryedon mauritanicus* only if it was lodged in the rare *A. macrostachya* seeds presenting a hole. In addition, as this polyvoltine bruchid was not found in the 2700 seeds collected on the other hosts noted by Decelle (1979), how could its generations be inferred to develop successively during the dry season, if its larvae could not develop in already utilized or still inhabited seeds?

Up to 3 *Bruchidius tougouriensis* can develop in one *Acacia macrostachya*, *Albizia lebbeck* or *Acacia dudgeoni* seed.

Caryedon pallidus is polyvoltine, though it has a solitary development, its pupation, which always occurs outside the seeds, and often after escaping the pods, may favour the infestation of new *Cassia tora* and *C. occidentalis* pods (a minimum 30 seeds per pod) eventhough some of their seeds have already been parasitized. The adults coming from *C. occidentalis* small seeds are miniaturized.

Polyvoltine *Caryedon serratus* reproduces all the year round after it has been collected on *Arachys hypogea* L. Several larvae can develop in the same seed. Its very rudimentary behaviour allows a similarly important quantity of eggs to be laid on glass surfaces, even in the absence of pods to stimulate the release (J. Appert). This behavioural plasticity can account for the infestation of *Arachys hypogea* pods, while all the hosts it uses in situ are caesalpiniaceae. All cocoons picked on *Cassia sieberiana* contained adults in imaginal diapause on 24th June; now Donaye and his colleagues (1966) have observed its diapause on *Acacia spirocarpa* and *A. tortilis* in Israel. According to Prevett,

Table 5. Host-plants of Bruchidea in Upper Volta.

Bruchidae	Host-plant		
Bruchidius dichrostachydis (Decelle. n.sp.ined)	*Dichrostachys glomerate* (Forsk) Chiov.	mimosacée	
Bruchidius luteopygus (Pic.)	*Acacia sieberiana* D.C.	mimosacée	
Bruchidius pennatae (Decelle n.sp.ined)	*Acacia pennata* (L.) Wild.ex Dell.	mimosacée	10 species with one host-plant
Bruchidius sieberianae (Fahr.)	*Acacia seyal* A.Ch.	mimosacée	
Bruchidius submaculatus (Fahr.)	*Albizia lebbeck* (L.) Benth.	mimosacée	
Bruchidius voltaicus (Decelle n.sp.ined)	*Acacia seyal* Del.	mimosacée	
Caryedon mauritanicus (Decelle n.sp.ined)	*Acacia macrostachya* Reichenb.ex Benth.	mimosacée	
Caryedon sahelicus (Decelle)	*Acacia seyal* Del.	mimosacée	
Tuberculobruchus natalensis (Pic.)	*Acacia sieberiana* D.C.	mimosacée	
Bruchidius chloroticus (Dalm.)	*Sesbania sesban* L. Merril	papilionacée	
Bruchidius chloroticus (Dalm.)	*Sesbania pachycarpa* D.C.	papilionacée	4 species with 2 host-plants
Callosobruchus maculatus (F.)	*Vigna unguiculata* (L.) Walp.	papilionacée	
Callosobruchus maculatus (F.)	*Voandzeia subterranea* (L.) D.C.	papilionacée	
Caryedon interstinctus (Fahr.)	*Acacia sieberiana* D.C.	mimosacée	
Caryedon interstinctus (Fahr.)	*Acacia nilotica* (L.)(Guill et Perr) Oktze	mimosacée	

Table 5 (*continued*)

Bruchidae	Host-plant		
Caryedon pallidus (01.)	*Cassia occidentalis* L.	cesalpin acée	
Caryedon pallidus (01.)	*Cassia tora* L.	cesalpin acée	
Bruchidius tougouriensis (Decelle n.sp.ined)	*Acacia dudgeoni* Craib ex Holl	mimosacée	1 species with 3 host-plants
Bruchidius tougouriensis (Decelle n.sp.ined)	*Acacia macrostachya* Reichenb ex Benth.	mimosacée	
Bruchidius tougouriensis (Decelle n.sp.ined)	*Albizia lebbeck* (L.) Benth	mimosacéc	
Caryedon serratus (01.)	*Tamarindus indica* L.	cesalpin acée	
Caryedon serratus (01.)	*Piliostigma reticulatum* (D.C.) Hochst	cesalpin acée	1 species with 4 host-plants
Caryedon serratus (01.)	*Piliostigma tonningii* (Schum) Milne Redhaed	cesalpin acée	
Caryedon serratus (01.)	*Cassia sieberiana* D.C.	cesalpin acée	

Table 6. Catching of adult bruchids.

Month	Number caught	Species caught	Location of catch
April	3	*Bruchidius albopubens* Pic	*Acacia dudgeoni* inflorescence
May	1	*Caryedon pallidus* 01	*Maerua angolensis* follicules (Cappasidae)
May	2	*Spermophagus* sp (*humilis*) J. Decelle	idem.
June	3	*Bruchidius dichrostachydis* J. Decelle	*Acacia macrostachya* pod near Dichrostachys *glomerata* plant
June	1	*Bruchidius Kiliwaensis* J. Decelle	
June	2	*Bruchidius senegalensis* Pic	*Acacia sieberiana* pod
June	1	*Callosobruchus maculatus* F.	*Acacia macrostachya* pod.
June	3	*Spermaphagus* sp (*humilis*) J. Decelle	*Acacia macrostachya* pod and inflorescence.

Sohi (1940) has noted it hibernates in the cocoons in India during 'winter' when it attacks *Tamarindus indica.* There *C. serratus* populations should be studied very precisely.

Caryedon interstinctus presents a diapause (an imaginal one?) in the cocoons formed in the litter or the soil, after escaping from the deciduous indehiscent pods of *Acacia nilotica* and *A. sieberiana.*

The legumes attacked by 3 *Caryedon* with an external pupation do not present an homogenous fructification behaviour; some have persistent dehiscent pods, others deciduous indehiscent ones . . . .

Caryedon interstinctus parasitizes *Acacia nilotica* and *A. sieberiana* which grow close to each other, with deciduous and dehiscent pods, and seeds the same size. But its infestation rates are respectively 0.04 and 0.15.

Bruchidius tougouriensis attacks three trees abonding in Saria, with seeds of a similar size: *Acacia dudgeoni*, *A. macrostachya* and *Albizia lebbeck.* The adults leave them at the same date, which is likely to imply synchronous ovipositions. It alone attacks *A. dudgeoni*, but it competes with *Caryedon mauritanicus* on *A. macrostachya*, and with *Bruchidius submaculatus* on *Albizia lebbeck.* Now, the infestation rates of these 3 plants by *B. tougouriensis* are respectively about 0.02, 1.3 and 0.04. Therefore, *the system of relationship between bruchids and legumes is not a binary one* according to the capacity to get through the allelochemical protection, with the alternative: presence or absence of attack. The possibilities of bruchids are modulated by many factors.

CONCLUDING REMARKS

The exceptionally homogenous geographical and ecological conditions prevailing in the Mossi land, and the synchronisation of pickings allowed to expect certain ecological characteristics related to adaptive values to be set in evidence, since modern works are full of descriptions of ecological 'strategies', 'discovered' in far more complex, variable and heterogenous ecosystems, following often less frequent pickings.

Janzen (1977) considers that the fall of big indehiscent fruits is a 'strategy' favouring their dissemination by small mammals; now the deciduousness of *Tamarindus indica* or *Parkia biglobosa* fruits complies with this hypothesis, but not that of the small pods with small seeds of *Dicrostachys glomerata*, or of the fine pods with equally small seeds of *Cassia tora.*

Nor can polyvoltinism or monovoltinism, solitary or gregarious development be inferred from the nature of fruits or the type of fructification.

While the characteristics of the host impose constraints to its utilization, various adaptative ways allow these constraints to be overcome; The ecological and ethological adaptation are also concerned by the convergent evolution of species. Jacob (1977) has recalled that Darwin had emphasized that circumstantial aspect of adaptations.

However a few general remarks can be made. There is no relation between the number of adults collected and the number of those actually present. Finding a bruchid on a pod does not mean this bruchid parasitizes that legume.

Most introduced species are not attacked by Bruchids. Lastly, while it can be assumed with Orians *et al* (1974) that there exists 'a correlation between the length of time a tissue is present, the kinds of functions it performs, and the type of antiherbivore defenses it has', yet the defence should not be reduced to its chemical aspect, however important the latter may be. For instance, after digging out, *Voandzeia subterranea* or *Arachys hypogea* pods are extremely vulnerable; *Cajanus cajan* seeds are heavily attacked once out of their pilose, sticky pods.

The observations on the importance of bruchid attacks entail another question: what is the importance of sexual reproduction in perennal legumes? They are rarely dehiscent, their mechanical dissemination (by ejection) or wind is hardly conceivable, and their germination in situ is very poor. Then is the dissemination of seeds by vertebrates to be considered paramount? Would then sexual reproduction exist only as an insurance against ecological accidents, as was suggested by Bradshaw? But then what can be the meaning of calculatings on the energetic yield of sexual reproduction in legumes? How would the evolution of legumes be determined by the selection pressures exercised by bruchids? Are the allelochemical substances of legume seeds important because of their sole protective role, or do not they rather operate as an overall protection against all aggressions including those of pathogens?

The biocenosis and trophic chain concepts are corollary to the coevolution concept. Therefore, the latter cannot be reduced to the evolutionary relationships between *one* plant and *one* organism. *It is indispensable to study cultivated plants in their original biocenoses to understand these relationships.*

REFERENCES

Appert, J., 1954. Bull. Agron. 13: 184–90.
Decelle, J., 1979. Insects of Saoudi Arabia, I.,

Donahaye, E., Navarro, S. & Calderon, M., 1966. Observations on the life cycle of *Caryedon gonagra* F. on its natural hosts in Israel, *Acacia spirocarpa* and *A. tortilis*. Trop. Sci., 8: 85–89.

Jacob, F., 1977. Evolution and tinkering. Science 196: (4295), 1161–1166.

Janzen, D.H., 1969. Seed-eaters versus seed size, number, toxicity and dispersal. Evolution 23 (1): 1–27.

Janzen, D.H., 1977. The interaction of seed predators and seed chemistry, pp. 415–428. In: Comportement des insectes et milieu trophique, Coll. Int. du C.N.R.S. 265: 493 pp.

Kamito, A., 1933. On the broad bean weevil introduced in Japan. Proc. Imp. Acad. Tokyo 9 (3): 137–39.

Kerharo, J. & Adams, J.G., 1971. La pharmacopée senegalaise traditionnelle: plantes toxiques et médicinales. Vigot, Paris.

Loukianovitch, F.K. & Ter. Minacian, M.E., 1957. Joukiternoyki (Bruchidae). Faune de l'U.R.S.S. 24 (1): 67, 1–210.

Orians, G.H., Gates, R.G. Rhoades, D.F. & Shultz, J.C., 1974. Producer-consumer interactions. Proc. 1st Int. Cong. Ecol., 213–217.

Prevett, P.F., 1966. Observations on the biology of six species of Bruchidaea (Coleoptera) with particular reference to the northern Nigeria. Entomologist's Monthly Magaz. 102: 1226–1228, 174–180.

Prevett, P.F., 1966. Observations on Biology in the genus *Caryedon* Schönherr (Coleoptera: Bruchidae) in Northern Nigeria, with a list of associated parasitic Hymenoptera. Proc. R. Ent. Soc. London. (A), 41: (1–3), 9–16.

Prevett, P.F., 1967. Notes on the biology, food plants and distribution of Nigerian Bruchidae (Coleoptera) with particular reference to the northern region. Bull. Ent. Soc. Nigeria 1: 3–6.

Sohi, G.S., 1940. *Pachymerus gonagra* as a store pest. Indian J. Ent. 2: 244.

Zeven, A.C. & Zhukovsky, P.M., 1975. Dictionary of cultivated plants and their centres of diversity. Pudoc, Wageningen, 219 pp.

Authors' address:
I.B.E.A.S. (ERA CNRS 328)
Université F. Rabelais
Parc de Grandmont
37200 Tours
France

ROLE OF ALLELOCHEMICS IN THE SPECIALISATION OF TROPHIC RELATIONS BETWEEN BRUCHIS AND LEGUMES

GERALD A. ROSENTHAL
(Lexington, Kentucky, U.S.A.)

At least 500 leguminous species are presently known that store L-canavanine (Bell *et al.*, 1978), a highly toxic and insecticidal non-protein amino acid bearing structural analogy to L-arginine.

$$H_2N - C(=NH) - NH - O - CH_2 - CH_2 - CH(NH_2)COOH$$

L-canavanine

Dioclea megacarpa, an inhabitant of the deciduous forests of Costa Rica, is distinctive in that canavanine accounts for up to 13% of the seed dry weight (Rosenthal *et al.*, 1976). So much toxic canavanine represents a formidable chemical barrier to a phytophagous insect but the bruchid beetle, *Caryedes brasiliensis* is unhampered in its aggressive attack on the seed. The developing larvae totally ravage the canavanine-laden cotyledonous tissues.

Studies of this seed predator have provided substantive insights into its biochemical adaptation to this normally poisonous seed. Ordinarily, direct consumption of significant canavanine-containing tissue creates a severe biological hazard since the arginyl-tRNA synthetase of non-producing species activates canavanine. This ultimately results in the synthesis of anomalous, canavanine-containing proteins. A major biochemical adaptation by this insect to the appreciable canavanine in its foodstuffs is avoidance of canavanyl protein production which is achieved by an arginyl-tRNA synthetase able to discriminate between canavanine and arginine. In this way, canavanine is not activated nor covalently linked to the growing polypeptide chain (Rosenthal *et al.*, 1976). This discriminatory ability represents a major biochemical adaptation of the bruchid beetle to this toxic food constituent.

Evaluation of the soluble non-protein amino acids of *Dioclea* seeds disclosed that canavanine contributes 96% of the total ninhydrin-positive material; other free seed amino acids are quantitatively insignificant. The gut of some bruchid beetle larvae exhibit limited proteolytic activity (Applebaum, 1964). This reflects a response of the larvae to the presence of proteolytic enzymes found in the seeds they consume and limits seed protein as a dietary amino acid source. In view of these factors, it seems quite possible that canavanine may function as a dietary nitrogen source for the developing insect.

Bruchid beetle larvae share with other insects demonstrative arginase activity which provides ornithine for their metabolism. It also creates the means to convert canavanine to canaline and urea. Urea cannot be assimilated directly but its hydrolysis to ammonia circumvents this metabolic impediment. Urease, required for this conversion, is rarely reported in insects and accounts of its occurrence have generally been discounted (Cochran, 1975); in fact, this insect is a prodigious urease producer (Rosenthal *et al.*, 1977). Formation of canaline and urea and then CO_2 and ammonia account for about three-quarters of the total insect catabolism of canavanine.

These observations suggest that this seed predator does not simply excrete, detoxify, or otherwise avoid this poisonous compound, but rather uses canavanine as a major dietary source of nitrogen. The bruchid beetle overcomes the disproportional sequestration of seed nitrogen into canavanine and the limited utility of seed storage protein by its sequential conversion to canaline and urea and then ammonia. Urease production permits nitrogen to be generated in a form that is fixed facilely into organic linkage. This represents a second significant biochemical adaptation resulting in conversion of canavanine to a beneficial nitrogenous resource.

It is interesting to speculate whether this is actually achieved by propagation of a biochemically capable symbiont or from information constituting part of the larval genome. While this is presently unknown, it is clear that the observed urease does not represent vitiation by ingested seed materials. The urease activity per mg soluble larval potein is even greater than that noted for comparable extracts of many canavanine-containing legumes which are among the richest sources of this enzyme to be found in the plant kingdom. In addition, the pupae are also a potent urease source and the larvae are known to purge their gut contents and curtail feeding activity as a requisite to pupation. Thus, it is wrcng to rationalize the observed urease activity on the basis of the minute amount of seed material which may be present in the gut.

Release of canavanine's stored nitrogen by urea formation also produces a stoichiometric yield of canaline, a highly toxic non-protein amino acid in its own right (see Rosenthal, 1978). This biochemical predicament is circumvented by reductive deamination of canaline to form homoserine and ammonia.

$$H_2N - C(=NH) - NH - O - CH_2 - CH_2 - CH(NH_2)CO_2H \rightarrow \quad \text{(i)}$$

L-canavanine

$$H_2N - O - CH_2 - CH_2 - CH(NH_2)CO_2H + H_2N - C(=O) - NH_2$$

L-canaline urea

$$H_2N - C(=O) - NH_2 \rightarrow CO_2 + 2NH_3 \quad \text{(ii)}$$

urea

$$H_2N - O - CH_2 - CH_2 - CH(NH_2)CO_2H + NADH + H^+ \rightarrow \quad \text{(iii)}$$

L-canaline

$$HO - CH_2 - CH_2 - CH(NH_2)CO_2H + NH_3 + NAD^+$$

L-homoserine

By conducting reactions i through iii, *C. brasiliensis* detoxifies canavanine plus canaline and provides homoserine and NH_3. Homoserine formation is itself important since it conserves amino acid carbon skeleton usable in other reactions (Rosenthal *et al.*, 1978).

The ability of this seed predator to reductively deaminate canaline not only effectively nullifies the toxin, but also more fully utilizes canavanine's stored nitrogen. It represents the third elucidated biochemical adaptation of this seed predator that permits efficient utilization of the seed nitrogen allocated to canavanine production and storage.

These findings support the conclusion that this bruchid beetle adapted to the toxic components of *Dioclea* by functioning biochemically as a canavanine synthesizing plant. This example of convergent evolution is suggested by three lines of evidence. First, the production of an insectan arginyl-tRNA synthetase able to discriminate between arginine and canavanine. Higher plants avoid non-protein amino acid autotoxicity with aminoacyl-tRNA synthetases able to discern between a protein amino acid and one that is not (Fowden & Lea, 1979). Second, sequential arginase- and urease-mediated hydrolysis that ultimately convert canavanine to canaline, CO_2, and NH_3. This biochemical sequence is also the principal catabolic pathway for mobilizing canavanine's nitrogen by legumes (Rosenthal, 1970). The final parallel is revealed by canaline's detoxification via conversion to homoserine. Canavanine synthesis in the jack bean plant, *Canavalia ensiformis*, proceeds by a cyclic reaction sequence involving canaline. Carbon may enter this cycle through transformation of homoserine to canaline. Thus, while the purpose of canaline and homoserine interconversion differ in the bruchid beetle from that of the legume, these non-protein amino acids share a vital role in the basic reactions constituting canavanine metabolism.

Simultaneous development of the three biochemical capacities described above represent an improbable occurrence, but one rendered more rational when viewed as a series of adaptive experiences. Perhaps a progenitor of *C. brasiliensis* was able to avoid canavanyl protein production as its extant offspring does but had not, as yet, adapted to canavanine or canaline. While the fitness of the early invaders of this food resource undoubtedly fell significantly – they survived! In time, the remnant population acquired the ability to manipulate canavanine. Canaline also came to be eventually degraded; it represented a means of detoxifying this poisonous natural product while enhancing the efficacy of carbon and nitrogen utilization from stored canavanine. Sequential acquisition of these biochemical capabilities continued to increase the overall biological fitness of this seed predator as it became better adjusted and more committed to its new food resource.

It is possible that the several accomodations inherent in the total adaptive process occurred over a very considerable time period. That is, early *D. megacarpa* may have stored little canavanine but seed predator pressure instigated an ever higher commitment of the seed nitrogen resource to canavanine production. The ever increasing concentration of canavanine gradually selected for those individuals able to accomodate the increasing level of this seed non-protein amino acid.

By ovipositioning solely on *D. megacarpa*, the female limits drastically the diversity, availability, and extent of its larval food supply. This immediately raises the question: why hasn't this predator moved to other canavanine-containing legumes of its habitat? Particularly since this herbivore has already invested in development of the metabolic mechanisms to fully exploit canavanine. Perhaps other canavanine-containing legumes store toxic compounds to which *C. brasiliensis* has not adapted or it may be unresponsive to the physical and/or chemical cues critical to locating and recognising potential canavanine-producing hosts. Being the only significant predator of this legume, *C. brasiliensis* most certainly benefits from unencumbered access to its food source; the usual competitive struggle for food resources need not concern this herbivore. In addition, it only needs a limited detoxification capability instead of a wide array of metabolically costly modes of response. While these factors represent cost-benefit items favoring the monophagous feeding habit of this seed predator, they severely minimize its ability to radiate into new food territories.

The interaction of *C. brasiliensis* with this leguminous species is an excellent example of biological compromise in which certain benefits from lack of competition or the need to maintain diversified detoxification strategies must be weighed against precarious dependency upon a single food supply. No plant is ever inviolate and *C. brasiliensis* has obviously succeeded in its coevolutionary interactions with this legume. Nevertheless, the toxic alleochemics of the seed afford it a remarkable degree of protection from herbivory and highlight the significant contribution of toxic secondary metabolites in the chemical defense of the plant.

ACKNOWLEDGEMENTS

The author gratefully acknowledges grant support from the National Institutes of Health, The National Science Foundation, and the University of Kentucky Research Foundation in the conduct of the author's work described in this review.

REFERENCES

Applebaum, S.W., 1964. J. Insect. Physiol. 10: 783.

Bell, E.A., Lackey, J.A. & Polhill, R.M., 1978. Biochem. Ecol. System. 6: 201.

Cochran, D.G., 1975. In: Insect Biochemistry and Function (D.J. Candy & B.A. Kilby, eds.), pp. 177–281. Chapman and Hall, London.

Fowden, L. & Lea, P.J., 1979. In: Herbivores: Their Interaction with Secondary Plant Metabolites (G.A. Rosenthal & D.H. Janzen, eds.), pp. 135–160. Academic Press, New York.

Rosenthal, G.A., 1970. Plant Physiol. 46: 273.

Rosenthal, G.A., 1978. Life Sci. 23: 93.

Rosenthal, G.A. Dahlman, D.L. & Janzen, D.H., 1976. Science 192: 256.

Rosenthal, G.A., Janzen, D.H. & Dahlman, D.L., 1977. Science 196: 658.

Rosenthal, G.A., Dahlman, D.L. & Janzen, D.H., 1978. Science 202: 528.

Author's address:
The Graduate Center for Toxicology and the
T.H. Morgan School of Biological Sciences
University of Kentucky
Lexington, Kentucky 40506
U.S.A.

FEEDING, LONGEVITY AND REPRODUCTION OF ADULTS OF *ACANTHOSCELIDES OBTECTUS* SAY IN LABORATORY CONDITIONS

B. LEROI
(*Tours, France*)

INTRODUCTION

In 1938, Larson and Fisher noted about *Acanthoscelides obtectus* Say and *Callosobruchus maculatus* F. that the feeding of adult Bruchidae had attracted little attention of researchers. It is still true at present. There is only one well-known exception, *Bruchus pisorum* L., whose females are known to be able to reproduce only after consuming some pollen from the flowers of their host-plant, *Pisum sativum* L. (Brindley, 1939; Pajni & Sood, 1975).

This lack of information is due on the one hand to the biology of bruchids in the fields and in their original ecosystems being poorly known, or even unknown, mainly for polyvoltine species; on the other hand, to the fact that the species most harmful to cultivated legumes have generally been considered like 'stored product' species, and in that case imaginal feeding did not have to play an important role, as most of these species have the reputation of being able to oviposit without feeding, owing to the important reserves accumulated in their fat body (Zacher, 1929). Lastly, the data concerning reproduction come from laboratory experiments in which the age of adults has not been taken into account, or from experiments bearing on the few days following their emergence without considering longevity capacities.

In the case of *A. obtectus*, very little is known about the adults' behaviour and their cycle in the original American equatorial ecosystems. The generations can apparently follow each other owing to the almost permanent presence of the known Phaseolinae hosts; but nothing is known about the adults' longevity, nor about their feeding capacities.

In the temperate areas of Europe or America, where the species is related to bean crops in the absence of wild Phaseolinae, very few data are available concerning the males' and females' longevity, their food sources, their flight capacities and for looking for new crops during the hot months, allowing them to be active outside stored grains. Some authors have assumed they might feed on the host-plant: on leaves, stipules, green pods . . . (Slingerland, 1893; Hawley, 1922 . . .) but the direct observation have not been made, or concern isolated data. De Luca (1966) reports adults visiting *Ranunculus* flowers, while Hase (1952) reports their presence in Germany on the flowers

Series Entomologica, Vol. 19: ed. by V. Labeyrie

of plants belonging to various families (Ranunculaceae, Rosaceae, Umbelliferae, Convolvulaceae, Compositae, Cruciferaceae). Zachariae (1958) lists nine plants visited by the adults 'to lick the flowers nectar'; he also records he has observed isolated individuals on the flowers of eleven other plants, though they were unable to get their nectar, and they might have been there 'accidentally'.

Similar observations have been made with other Bruchidae, such as *Callosobruchus maculatus* on cotton flowers (Ashmead, 1894), *Callosobruchus chinensis* on *Cajanus cajan* flowers (Williams, 1977), or *A. pauperculus* (Le Conte) on various flowers (Johnson, 1977). Personal observations bear especially on adults of *Megacerus* sp. observed on *Convolvulus arvensis* flowers in Putaendo (Chile), and on *Mimosestes nubigens* Motsch feeding on the pollen of a compositae growing in large number in the vicinity of the *Acacia farnesiana* hosts in the area of Durango (Mexico).

But I have never been able to observe *A. obtectus* adults feeding on *Phaseolus vulgaris*, or *Phaseolus coccineus*, or other wild Phaseolinae flowers in Mexico, nor on the flowers of cultivated beans (France, Chile) or on the flowers growing near wild or cultivated beans (France, Mexico, Chile). The only active adults observed were females on ripe bean pods at the time of oviposition or a few days before on bean leaves or stalks, though a typical feeding behaviour could not be noticed. In Mexico, adults have been observed during the dry season, sheltering in dry leaves of *Phaseolus* or of other plants (Biemont & Bonet, 1980), though a feeding behaviour could not be noticed during the periods of activity.

Indeed in the agrosystems that have been studied until now in France adult population of *A. obtectus* always seem to be small, which does not facilitate observations. Our failures may also have been caused by a fundamental mistake which consists in looking for these adults on the host-plants or on the neighbouring ones, whereas feeding might occur in other areas or other plant layers, taking into account the flight capacities we do not know.

We are reduced to assuming, as Zachariae did (1959) that *A. obtectus* adults can have water and food in the fields. By analogy with other bruchids or other phytophagous Coleoptera, it is assumed this feeding must be made on nectars and pollens preferably, possibly on certain plant tissues or fungi.

In stored beans, the data on longevity are just as unreliable, and the possibilities of feeding are considered non-existent: the reserves in the fat body are considered as being sufficient to ensure reproduction. However, Sinha (1971) has noticed that *A. obtectus* adults could ingest fungi growing in stores, though the latter cannot be positively stated to play a role in feeding.

EXAMINATION OF THE DIGESTIVE TRACT OF *ACANTHOSCELIDES OBTECTUS* ADULTS

The examination of the contents of the digestive tract has not yielded any information as yet that could direct research in nature. The insects collected in the fields apparently had ingested water or liquids. But no plant debris, no grains of pollen could be identified precisely. After a few tests of feeding on pollens, it seems the exine often gets broken by the adult's buccal parts, which makes it difficult or impossible to identify pollen grains in the digestive tract.

The enzymatic range has been studied with the Api-Zym method (Auxotab), which allows seventy enzymatic activities to be detected rapidly, and has already been utilized for insect digestive tracts (Plantevin & Nardon, 1972). This method has allowed a large number of enzymatic activities to be detected in *A. obtectus* adults, but no important differences were found resulting from their age or food. The particularly important activities of acid phosphatase, of β-glycosidases, of arabinosidase, and of a dehydrogenase have been shown, but they are activities which are usually to be found in phytophagous insects and nothing in the range of detected enzymatic activities allows to show a particular apparatus allowing research to be directed towards such or such sort of food.

EXPERIMENTAL STUDY OF THE INFLUENCE OF IMAGINAL FEEDING IN *ACANTHOSCELIDES OBTECTUS*

Larson & Fisher (1938), then Zachariae (1959) have noted that adults fed with sugar solutions or honeyed water present a higher longevity and generally lay more eggs in comparison to unfed adults. That is what we wanted to show precisely in the laboratory, by using types of diet almost similar to what can be found in nature (water, sugars, pollens . . .), by examining the individuals performances and placing the insects in various artifical situations not unlike certain ecological situations.

The experiments were made with adults coming from a strain maintained in the laboratory, corresponding to the offspring of females selected for their ovipositing only in the presence of beans. Egg and larval development was made at 27 °C. The adults used (about 50 in each serie) were placed individually in Petri boxes immediately after emergence. The fecundated females were inseminated on the 5th day of imaginal life and isolated after mating. Powder or homogenized food was provided in small saucers, water and liquid food on cotton-wool in these saucers, they were regularly changed and permanently provided (unless otherwise stated). In the series with beans, those were supplied by three and changed after each counting of laid eggs. When determining longevity, the insects were examined daily. For the reproductive records of females, the eggs laid (fecundity) and oocytes in retention were counted at the end of experiment or when each adult died: ovarian production corresponds to the sum of the two, as there is no oocyte retention in the lateral oviducts (Labeyrie, 1960).

The experiments were made either at 27 °C (RH = 60–70%, permanent darkness) or with a 12 h thermoperiod at 13 °C dark, and 12 h at 21 °C light (RH = 60–80%).

Influence of imaginal feeding upon the longevity of virgin adults

a) At 27 °C, a comparison was made of the influence of the following diets: powder sugars or a 20% sugar solution (saccharose, glucose, fructose), honeyed water, homogenized pollen pellets (with water), a mixture of honey and homogenized pollen (with water).

The results (Table 1) as compared with those of unfed adults or supplied with water only, showed, when food was provided permanently, that:

- the mean longevity of females is higher than that of males.
- longevity is remarkable enough in the absence of food.
- water alone does not augment longevity, but it is absolutely necessary for survival: food powder alone is not sufficient (except saccharose which is naturally very absorbent).
- the sugar solutions used in the experiments caused longevity to increase significantly; honeyed water also yielded good results, but significantly poorer than sugar solutions.
- the results yielded by pollen powder and the honey + pollen mixture were quite similar to those yielded by sugars.

These data bear out Larson & Fishers observations (1938) on the role played by sugars in *A. obtectus* longevity; this role known in many insects, has also been noted in *Bruchidius alfieri* Pic. and *B. trifolii* (Motsch.)(Hafez & Osman, 1956), or *Callosobruchus maculatus* (Larson & Fisher, 1924, 1938). Variability is high for all diets (Leroi, 1978).

Feeding must be permanent or regular to ensure survival: sugars or pollen provided in the first ten days of imaginal life were not sufficient to ensure a high longevity (series K to M, Table 1).

b) At 13 °–21 °C, the results confirmed the importance of sugars in imaginal feeding, though the longevities we obtained were higher than at 27 °C (Series N and O, Table 1). Some females live one year, and even longer

Table 1. Longevity of virgin adults of *A. obtectus* with different imaginal foods (27 °C or 13°/21°C).

	Series	Food	Nb	♂ Average longevity (days)	Nb	♀ Average longevity (days)
	A	Nothing	50	16.1 ± 0.8	51	22.5 ± 1.7
	B	Water	50	18.8 ± 1.3	48	24.9 ± 1.9
	C	Saccharose powder	49	25.6 ± 1.7	49	26.7 ± 2.0
	D	Glucose powder	49	16.8 ± 0.9	49	23.7 ± 1.3
27 °C	E	Saccharose solution	47	90.4 ± 12.1	48	152.2 ± 22.2
	F	Glucose solution	50	103.1 ± 10.2	48	172.0 ± 15.0
	G	Fructose solution	50	107.5 ± 14.0	48	171.9 ± 20.9
	H	Honeyed water	48	63.1 ± 6.1	50	85.8 ± 9.9
	I	Water + pollen	48	99.6 ± 15.4	45	170.4 ± 25.9
	J	Water + honey + pollen	49	100.1 ± 16.0	45	130.0 ± 26.9
	K	Saccharose sol. (10 days)	50	23.1 ± 1.1	49	29.4 ± 2.2
27 °C	L	Saccharose sol. (10 days)	51	24.1 ± 1.8	48	30.8 ± 2.3
	M	Water + pollen (10 days)	50	20.8 ± 0.9	50	28.8 ± 1.6
	N	Nothing	50	34.2 ± 1.8	50	50.2 ± 3.8
13 °/21 °C	O	Saccharose solution	46	157.8 ± 20.8	51	264.6 ± 27.6
	COL	Saccharose solution	51	191.2 ± 21.7	50	265.1 ± 27.8

when fed with a saccharose solution. Which was found again in adults of a strain coming from Columbia and maintained in the same conditions (series COL).

Influence of imaginal feeding on the ovarian production of virgin females

a) At 27 °C, the results shown (Table 2) concern the females of series A to J, maintained without beans:

- Imaginal feeding is not necessary to obtain an ovarian production. Up to 55 oocytes were counted in females of series A. This corresponds to the basic production, made only from the fat body. It is this production, together with the longevity of unfed adults which led to the assumption that the adults never feed.
- Sugar solutions, honeyed water, pollen, and especially the honey + pollen mixture, when permanently provided, stimulated ovarian production. Let us stress again, in all cases, a highly variable responsiveness of females to that stimulation (Fig. 1).

Table 2. Ovarian production of virgin females of *A. obtectus* with different foods (27 °C, without beans).

Series	Food	Average ovarian production
A	Nothing	31.0 ± 3.3
B	Water	26.9 ± 4.9
C	Saccharose powder	32.2 ± 4.3
D	Glucose powder	31.7 ± 3.3
E	Saccharose solution	36.4 ± 4.7
F	Glucose solution	38.0 ± 3.5
G	Fructose solution	43.0 ± 3.9
H	Honeyed water	37.9 ± 3.4
I	Water + pollen	44.0 ± 6.6
J	Water + honey + pollen	51.1 ± 6.5

Significative difference: ABCD, EFGHI, J.

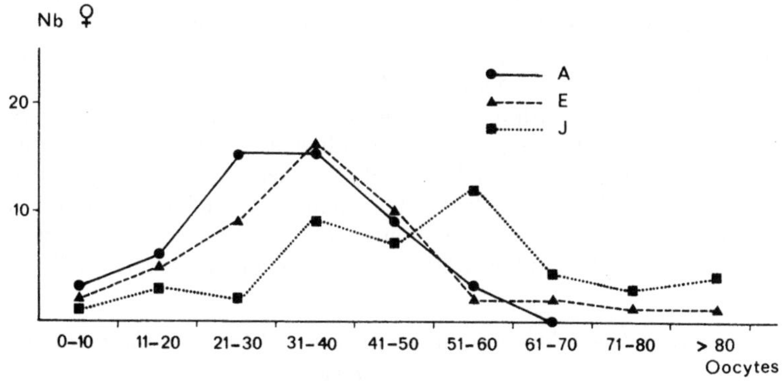

Fig. 1. Ovarian production of virgin females of *A. obtectus* unfed (series A), fed saccharose solution (series E) or fed with water + honey + pollen (series J).

– The analysis of individual records indicated there was no connection between ovarian production, oocyte retention, and longevity (Leroi, 1978).

b) At 13 °–21 °C, the stimulation of ovarian production by imaginal feeding was also considerable, but the performances were lower than at 27 °C. For example, a mean 30 oocytes were obtained with a saccharose solution, and only 16 without food.

c) In the females of the Columbian strain, the comparative study of the hemolymph of females fed on honeyed water or on honeyed water + pollen, made with the Laurell technique (1966) showed the supply of pollen proteins resulted in an increased production of vitellin proteins, and an increased quantity of oocytes which reach maturity (Huignard & Leroi, unpublished).

Influence of imaginal feeding on the longevity and ovarian production of fecundated females

The results are shown in Table 3 and Fig. 2.

a) At 27 °C, females deprived of food and bean grains (series P) presented a longevity comparable to that of unfed virgin females. Their similar ovarian production showed that, in that strain, fecundation does not cause stimulation, but this does not correspond to a general case: fecundity is generally stimulated by insemination.

b) At 27 °C, without any food but in the presence of bean grains (series Q), the mean ovarian production rose up to 72 oocytes, as a consequence of the stimulation by the host plant (Huignard, 1970). On the other hand, longevity decreased ($\overline{m} = 18$ days), as a consequence of the metabolic spendings effected for reproduction.

c) At 27 °C, with a saccharose solution but in the absence of beans (series R) a longevity comparable to that of virgin females provided with the same food ($\overline{m} = 173$ days) was found again, along with a considerable variability. The mean ovarian production was 38 oocytes, 80% of them being kept in retention (females selected for their ovipositing in the presence of beans).

d) At 27 °C, females fed with a saccharose solution and pollen powder, placed in the permanent presence of beans (series S) reached an average ovarian production of 226 oocytes, 96% of which were laid.

e) At 13 °–21 °C, with a saccharose solution, pollen, and in the presence of beans (series T), the records were even higher: a mean oocyte production of 280, of which 96% were laid. Actually as in series S, a very high variability could be observed (Fig. 2). But yet no female showed a production below 65 oocytes (a maximum of 58 without food), while some produced over 400 and even 500 oocytes, and over 80% of the oviposited eggs were fertile.

Influence of delayed fecundation and presence of beans on the reproduction performances

The experiments were conducted at 27 °C, with females supplied with a saccharose solution. They enabled to know if females which cannot lay their

Table 3. Ovarian production of inseminated females of *A. obtectus*, with different conditions of temperature, food and presence of beans.

Temperatures	Series	Food	Beans	Nombre femelles	Ovarian production (% egg laying)
27 °C	P	Nothing	—	49	27.8 ± 4.0 (24%)
	Q	Nothing	+	47	71.8 ± 4.9 (97%)
	R	Saccharose solution	—	49	37.7 ± 5.0 (20%)
	S	Sacc. sol. + pollen	+	44	225.9 ± 31.9 (96%)
13°/21 °C	T	Sacc. sol. + pollen	+	53	280.1 ± 34.6 (96%)

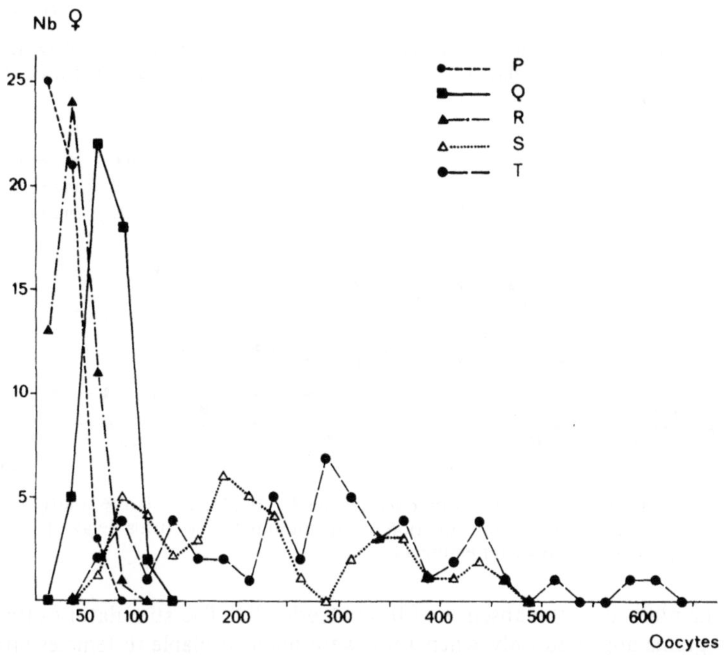

Fig. 2. Ovarian production of *A. obtectus* fecundated females: unfed without bean grains (P); unfed with bean grains (Q); fed saccharose solution, without bean grains (R); fed saccharose solution + pollen, with bean grains (S, T). Series P, Q, R, S: 27 °C; Series T: 13/21 °C.

eggs or (and) fecundated belatedly can retain their high reproductive capacities, and if the stimulations related to the presence of the host are still operative.

a) Females fecundated on the 5th day were placed in the presence of bean grains only after 20 days of imaginal life (series V), 50 days (series W) or 100 days (series Z). The results bear on the 10 days following the introduction of grains (Fig. 3).

The ovarian productions were not significantly different in series V, W, and TE (control females with beans at fecundation) much lower in seriez Z, whose females' production was comparable to that of TT females fecundated

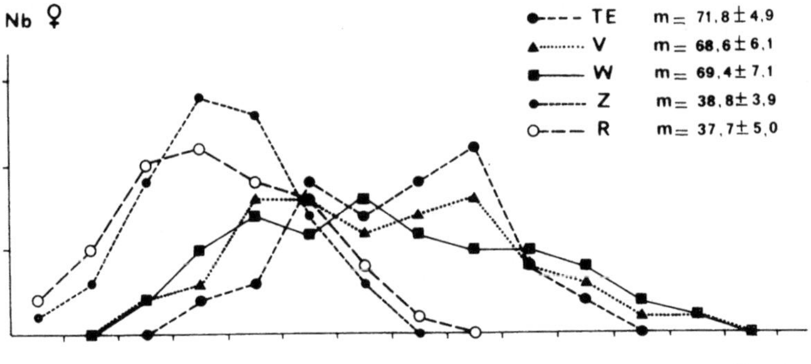

Fig. 3. Ovarian production of fecundated females of *A. obtectus* fed saccharose solution at 27 °C. Fecundation on day 5. Introduction of bean grains on day 5 (TE), 20 (V), 50 (W) or 100 (Z). No bean grains in series R.

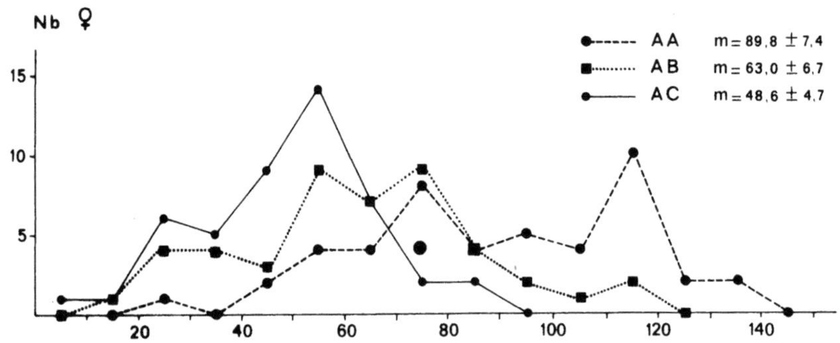

Fig. 4. Ovarian production of *A. obtectus* females (27 °C; fed saccharose solution) when fecundation and bean grains introduction occur simultaneously with a delay of 20 (AA), 50 (AB) or 100 (AC) days after emergence.

but maintained in the absence of beans (series R). The stimulatory effect of bean grains appeared only when they were made available to females tardily; in the conditions of experiment we have indicated, this stimulation would decrease progressively in females after 50 days of imaginal life. The retention of oocytes increased with the length of time which elapsed between fecundation and possible oviposition: from 3% in TE control females up to 75% in the females of series Z.

b) The influence of a delayed fecundation simultaneously with a delayed fecundation simultaneously with a delayed presence of grains was experimented: 20 days (series AA), 50 days (series AB), 100 days (series AC).

90 to 95% females mated on the 20th day, a proportion similar to that observed on the 5th day, and 85 to 90% on the 50th and the 100th day. The results on fecundity (Fig. 4) are to be compared with the above ones (Fig. 3).

The females in series AB had an ovarian production quite similar to that of females in series TE, V, and W. Those in series AC showed a poorer pro-

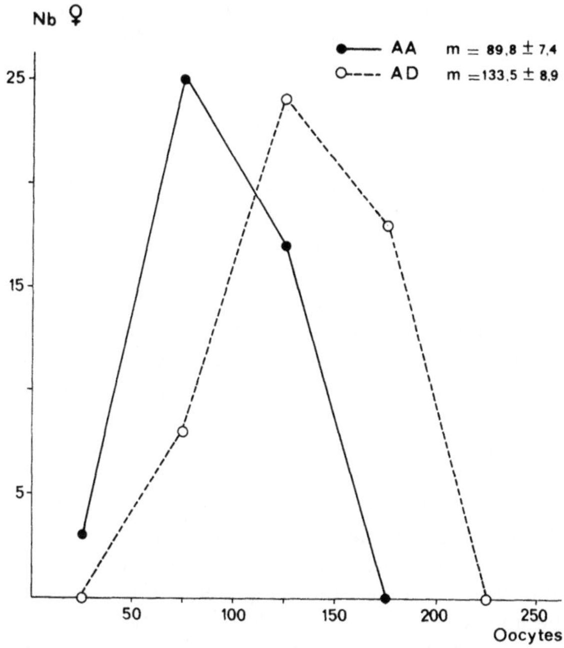

Fig. 5. Influence of alimentation on ovarian production of *A. obtectus* females fecundated on 5th day and supplied with bean grains on 20th day: Saccharose solution (AA) or saccharose solution + pollen (AD); 27 °C.

duction, though it was higher than that of females of the same age in series Z. The females in series AA showed a production significantly higher than that of females in the other series, due to a very high production in a few females. The oviposition rate was higher than in females of the same age in series V, W and Z, but it also decreased with ageing: 90% (AA), 80% (AB), 75% (AC). Fertility remained high: 90% (AA) to 75% (AC).

Thus the combined action of stimulatory factors (fecundation, presence of host) together with the action of feeding which allows a good longevity, was higher when these factors operated simultaneously. It is particularly interesting to note that this combined stimulation was highest about the 20th–30th days, that is when it occurred after the formation of the stock of oocytes formed out of larval reserves, a formation which is completed in that strain, towards the 15th day of imaginal life (Huignard, 1970). This is to be compared to Labeyrie & Maison's observations (1953) in South-west France: about 15 days elapse between the time weevils leave the granaries and the time they oviposit on beans in the fields, even though ripe pods are available when they fly off. This might be caused by a necessary imaginal feeding previous to oviposition.

c) When delaying simultaneously fecundation and the supply of bean grains to the 20th day went along with providing a mixed sugar and protein food (series AD: saccharose solution + pollen powder), the role played by protein food in the maturing of a complementary stock of oocytes could be noted once more. All the females were stimulated (Fig. 5).

DISCUSSION

The numerical data brought up here (longevity, ovarian production, oviposition) are valid in precise experimental situations and for one strain of *A. obtectus.* As they are, they should not be extended to other situations, and the multiplicity of stimulatory factors which can operate for quite a long period of life shows it is necessary, for each given ecological situation, to know thoroughly the ecological parameters and their variations in time, even within apparently homogenous populations, the considerable variability of the individuals' responsiveness to the factors which stimulate longevity and fecundity is to be noticed.

As the potentialities of survival of adults of *A. obtectus* are important, and the reproductive capacities of aged adults are high in favourable ecological conditions, it is obviously necessary to know thoroughly for each given situation the characteristics of the environment that are likely to cause fluctuations in fecundity, therefore in the level of larval populations which destroy the grains: the climate, but also the components of the vegetable environment in which the adults can feed. Small adult populations at the start are likely to cause a heavy infestation of crops, especially if the flight capacities are high.

This emphasizes how desirable it is now that the action of these factors in the fields should be appraised. We should study precisely the capacities for spreading of bruchids to look for food, host-plants and shelters. We must know if the adults feed regularly, where and when they do it and we must stop sticking to a few scattered observations. This should be carried out in various environments, according to the characteristics of latitude of the climate, and of the components of the ecosystems. And it seems absolutely necessary to extend these studies, in the laboratory and in natural environments, to the other Bruchids which attack man-consumed Leguminosae.

REFERENCES

Ashmead, W.H., 1894. Notes on cotton insects found in Mississipi. Insect Life 7: 240–247.

Biemont, J.C., & Bonet, A., 1980. Preliminary observations of bean weevil populations from the *Acanthoscelides obtectus* Say group living on wild or subspontaneous *Phaseolus vulgaris* L. and *Phaseolus coccineus* L. and on *Phaseolus vulgaris* L. cultivated in the Tepoztlan region state of Morelos-Mexico, this volume.

Brindley, T.A., 1939. Biology and control of the pea weevil in the Palouse area of Idaho and Washington. J. Econ. Entomol. 32: 21–24.

Hafez, M. & Osman, F.H., 1956. Biological studies on *Bruchidius trifolii* and *Bruchidius alfieri* in Egypt. Bull. Soc. Entomol. Egypte 40: 231–277.

Hase, A., 1952. Die Einbürgerung des Speisebohnenkäfers als Freilandschädling in Deutschland. Trans. 9th Int. Congr. Entomol. Amsterdam 1951. I: 666–668.

Hawley, I.M., 1922. Insect and other animal pest injurious to field beans in New York. N.Y. Agron. Expt. Stn. Mem. 55: 945–1037.

Huignard, J., 1970. Analyse expérimentale de certains stimuli externes influencant l'ovogenèse chez *Acanthoscelides obtectus* Say. Coll. Int. CNRS L'influence des stimuli externes sur la gamétogenèse des insectes, Tours 1969, pp. 357–380.
Johnson, C.D., 1977. Notes on the host plants and distribution of *Acanthoscelides pauperculus* (Coleopt. Bruchidae). Pan-Pacific Entomol. 53: 303–304.
Labeyrie, V., 1960. Action de la présence de graines de haricots sur l'ovogenèse d'*Acanthoscelides obtectus*. C.R. Acad. Sci. Paris 250: 2626–2628.
Labeyrie, V. & Maison, P., 1953. Observations préliminaires sur le comportement de la bruche du haricot *Acanthoscelides obtectus* Say dans la nature. Actes Congr. Luxembourg, Assoc. Fr. Avancement Sci. 73: 473–475.
Larson, A.O. & Fisher, C.K., 1924. Longevity and fecundity of *Bruchus quadrimaculatus* Fab. as influenced by different foods. J. Agric. Res. 29: 297–305.
Larson, A.O. & Fisher, C.K., 1938. The bean weevil and the southern cowpea weevil in California. USDA Tech. Bull. 593: 71 pp.
Laurell, C.B., 1966. Quantitative estimation of proteins by electrophoresis in agarose gel containing antibodies. Anal. Biochem. 15: 45.
Leroi, B. 1978. Alimentation des adultes d'*Acanthoscelides obtectus* Say (Coléoptère, Bruchidae): influence sur la longévité et la production ovarienne des individus vierges. Ann. Zool. Ecol. Anim. 10: 559–567.
Luca (De), Y., 1966. Alimentation imaginale des Bruchides (Col.) Parasitica 22: 26–54.
Pajni, H.R. & Sood, S., 1975. Effect of peapollen feeding on maturation and copulation in the beetle *Bruchus pisorum* L. Ind. J. Exp. Biol. 13: 202–203.
Plantevin, G. & Nardon, P., 1972. Utilisation d'une microméthode de détection (Auxotab) pour la recherche qualitative d'activités enzymatiques dans les tissus d'insectes. Tube digestif. Ann. Zool. Ecol. Anim. 4: 229–248.
Sinha, R.N., 1971. Fungus as food for some stored-product insects. J. Econ. Entomol. 64: 3–6.
Slingerland, M.V., 1893. Some observations upon two species of *Bruchus*. Psyche 6: 445–449.
Williams, I.H., 1977. Behaviour of insects foraging on pigeon pea (*Cajanus cajan*) in India. Trop. Agric. 54: 353–363.
Zachariae, G., 1958. Das Verhalten des Speisebohnenkäfers *Acanthoscelides obtectus* Say (Coleoptera: Bruchidae) im Freien in Norddeutschland. Z. ang. Entomol. 43: 345–365.
Zachariae, G., 1959. Kann sich der Speisebohnenköfers *Acanthoscelides obtectus* Say als Freilandschädling in Norddeutschland einbürgern?. Z. ang. Entomol. 45: 225–267.
Zacher, F., 1929. Nahrungsanswahl und Biologie der Samenkäfer. Verhandl. Deutsch. Ges. ang. Entomol. 1928: 55–62.

Author's address:
I.B.E.A.S. (ERA CNRS 328)
Université F. Rabelais
Parc Grandmont
37200 Tours
France

SOME DATA ON THE REPRODUCTIVE ACTIVITY OF *ZABROTES SUBFASCIATUS* IN THE LABORATORY

D. PIERRE & M. PIMBERT
(*Tours, France*)

INTRODUCTION

Zabrotes subfasciatus has been recorded in most tropical regions of Latin America and probably originated in one of these regions (Bondar, 1937). Like *Acanthoscelides obtectus* and *Callosobruchus maculatus* (Balachowsky, 1962) this insect can reproduce both in the field – before harvest of Phaseolinae seeds – and in storage. Jarry & Bonet (personal communication) have indeed observed that in Mexico *Zabrotes subfasciatus* reproduces in the field inside mature, dehiscent *Phaseolus vulgaris* pods and then on harvested seeds.

Although many studies of the biology of *Zabrotes subfasciatus* have been carried out on seeds in storage conditions (Steffan, 1945; Mendes Ferreira, 1960; Howe & Currie, 1964; Zacher, 1930; Carvalho & Rossetto, 1968), no study of this insect in the field had been made prior to Jarry & Bonet's observations.

Zabrotes subfasciatus has also been reported attacking stored Phaseolinae in tropical regions of Asia and Africa (Davies, 1977; Singh *et al.*, 1979). However, there are few clear records of this insect attacking these plants' seeds in the field.

Zabrotes subfasciatus females would be able to reproduce on and larvae could develop in seeds of several Phaseolinae and Viciae playing important roles in human nutrition (Ackroyd & Daughty, 1964): *Phaseolus vulgaris Phaseolus coccineus*, *Phaseolus lunatus*, *Phaseolus acutifolius*, *Vigna mungo*, *Vigna unguiculata*, *Vigna angularis*, *Lablab purpureus*, *Cajanus indicus*, *Cicer sp.*, *Vicia faba*, (Mendes Ferreira, 1960; Zacher 1930; Davies, 1972; Briwell 1918, Lepesme, 1944; Binaghi, 1955).

However, the behaviour of *Zabrotes subfasciatus* in presence of these plants cannot be predicted from these studies since they were carried out under artificial conditions. Ecological studies are therefore required in order to understand the life cycle of this insect as well as contamination conditions in the field, before the seed harvest.

In this preliminary study, the reproductive activity of *Zabrotes subfasciatus* on pods and seeds of *Phaseolus vulgaris* was examined under laboratory conditions approximating as closely as possible thermo and

Series Entomologica, Vol. 19: ed. by V. Labeyrie

photoperiodic regimes of the Costa Rican region from which the strain used in our experiments came from.

MATERIALS AND METHODS

Characteristics of the insect. Zabrotes subfasciatus shows a pronounced sexual dimorphism. The male is smaller (length 1.5 to 2.5 mm) than the female (2 to 3 mm). The latter's elytra are dark in colour with white patches usually merging into a white medio-lateral band. The male's elytra are uniformly light brown.

Females lay their hemispheric eggs on seeds of the host plant. The larva perforates the seed coat directly beneath the egg. Its development is complete after 4 larval stages. The larva of the 4 th instar cuts out an operculum (Steffan, 1945) which is used by the adult when emerging from the seed. Adults are sexually mature as soon as they emerge from the seed and live for 20 to 50 days.

Rearing conditions. The strain used came from contaminated *Phaseolus vulgaris* seeds collected in Jaco (Ouest coast of Costa Rica 9° Latitude North, altitude < 50 m) by Labeyrie in 1978. Experimental populations were maintained under the following conditions:

– Photoperiod 12 LL/12 LD
– Thermoperiod 33 °CL/23 °C.D.
– Hygroperiod 70%L/75–80%D

Isolated adult insects were offered *Phaseolus vulgaris* seeds (Soisson variety – harvested in 1979 – stored at 5 °C) and a trough regularly filled up with a sugar solution. The influence of insemination as well as the influence of pods and seeds on the reproductive activity of females was studied. All experiments lasted 30 days. At the end of experiments the fecundity (i.e., the number of eggs laid) was estimated for each female. Females were then dissected in order to calculate ovarian production (i.e., the sum of the fecdity and the number of eggs retained in lateral oviducts) and to detect under the microscope any spermatozoa present in the spermatheca.

RESULTS

1. *Reproductive activity of virgin isolated females* (*control*)

Most isolated virgin females did not lay eggs; 12 out of 35 laid a few eggs (1 to 2) which were scattered in the petri dish. Females produced a small number of ovocytes which accumulated in the lateral oviducts (Table 1) and were retained there without being resorbed.

Table 1. Reproductive activity of virgin females.

	Virgin females (controls)	Virgin females in presence of closed pods	Virgin females in presence of seeds
Total number of females	35	24	24
Fecundity*	0.45 ± 0.12	0.33 ± 0.12	35.44 ± 2.95
Number of females which did not lay eggs.	23	17	0
Ovarian production*	8.62 ± 0.92	18.83 ± 1.26[1]	46.25 ± 2.55[1]

*Mean ± SEM
[1] Significantly different from control ($P < 0.001$)

2. *Reproductive activity of virgin females in presence of closed pods of* P. vulgaris

Fecundity was very low when females were in presence of pods. Only 7 out out of 24 females laid a few eggs (1 to 2). However, ovarian production was slightly higher than the controls'. Pod presence does therefore not induce egg-laying but does slightly stimulate oogenesis (Table 1, Fig 1).

3. *Reproductive activity of virgin females in presence of seeds of* P. vulgaris

Seed presence stimulates oogenesis since ovarian production was significantly higher than the controls (Table 1, Fig. 1).

All females laid eggs which were glued onto the seeds. Egg laying activity only began after 15 days presence with seeds. Females did not lay all the eggs they produced as can be seen from the relation between ovarian production and fecundity (Fig. 2). However, the variation in these results differs from that obtained for virgin females of *Acanthoscelides obtectus* in presence of seeds (Huignard, 1970).

4. *Reproductive activity of isolated inseminated females*

Under these conditions, copulation frequency was very low (0.26). During copulation, spermatozoa and male secretions were emitted and introduced in the female's bursa copulatrix without spermatorphore formation (Pierre, 1979).

The number of eggs laid was low and not significantly different from that of virgin females placed under similar experimental conditions (Table 2).

In comparison with isolated virgin females, ovarian production was slightly more stimulated in this case.

Table 2. Reproductive activity of inseminated females.

	Virgin females (control) (1) (n = 35)	Isolated inseminated females	Inseminated females in presence of closed pods	Inseminated females in (1) presence of open pods	Inseminated females with seeds
Number of pairs	–	45	45	40	45
Copulation frequency (number of copulations)	–	0.26 (12)	0.20 (9)	0.92 (37)	0.86 (39)
Fecundity*	0.45 ± 0.12	0.25 ± 0.25	1.25 ± 1.25	56.67 ± 5.09	47.70 ± 3.98
Number of females which did not lay eggs.	23	11	8	0	0
Ovarian production*	8.62 ± 0.82	14.25 ± 1.44[1]	19.5 ± 2.75[2]	57.64 ± 5.04[3]	60.76 ± 2.92[4]

*Mean ± SEM

[1] Significantly different from control ($P < 0.01$).

[2, 3] and [4] significantly different from control ($P < 0.001$).

(1) 1 and 2: not significantly different.
3 and 4: not significantly different.

5. *Reproductive activity of inseminated females in presence of non-dehiscent* (Phaseolus vulgaris) *pods*

The frequency of copulation was low (0.20). Pods would therefore have no stimulant effect on copulation.

Non-dehiscent pods had no stimulant effect on fecundity or ovarian production (only 1 out of 9 inseminated females laid eggs). The latter was not significantly different with the ovarian production of virgin females placed in similar conditions or that of isolated inseminated females (Table 2, Fig. 3).

6. *Reproductive activity of inseminated females in presence of open pods*

A hole (1.5 × 2 mm) was made at the end of each pod at the level of the first seed (the seed furthest away from the peduncle).

Copulation frequency was very high (0.86–0.93). Insects explored the pod with the help of their antennae and palps for a considerable length of time, entered the pod via the hole, copulated inside and remained there for most of their life.

Females laid their eggs on all seeds accessible in the pod and eggs were distributed in a regular manner among seeds. A few eggs were laid on the internal walls of the pods (Table 2, Fig. 3).

Contact with seeds inside the pods stimulated oogenesis and induced egg laying activity.

7. *Reproductive activity of inseminated females in presence of seeds of* · P. vulgaris

Under these conditions, 86% of the females copulated. Seed presence stimulated oogenesis and induced egg laying. Results were very similar to those observed for inseminated females in presence of open pods.

As can be seen from the relationship between fecundity and ovarian production (Fig. 4), females did not lay all the ocovytes they produced under these experimental conditions. Moreover, induction of ovarian production could result from an interaction between the respective influences of seed and insemination. Ovarian production of virgin females in presence of seeds (46.2) + ovarian production of inseminated females without seeds (14.2) approximately equals the ovarian production of inseminated females + seeds (60.7).

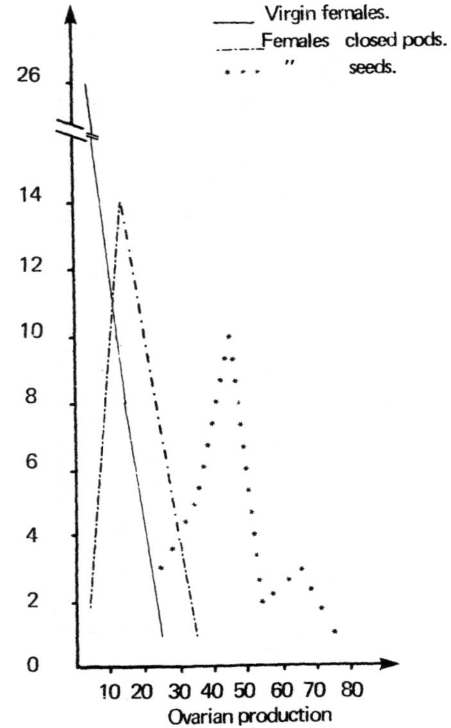

Fig. 1. Distribution of ovarian production of virgin females in presence of pods or bean seeds.

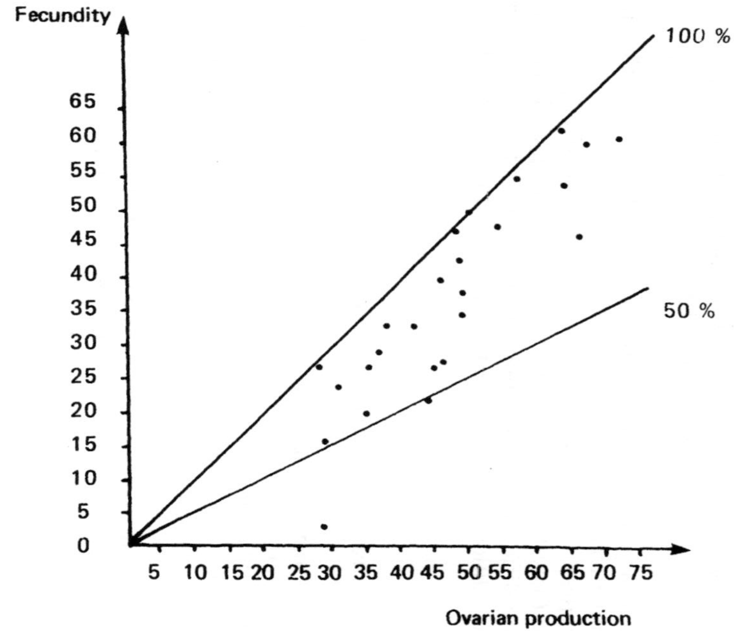

Fig. 2. Relation between fecundity and ovarian production of virgin females. The straight line (100%) indicates that fecundity equals ovarian production. The straight line (50%) indicates that fecundity equals half of ovarian production.

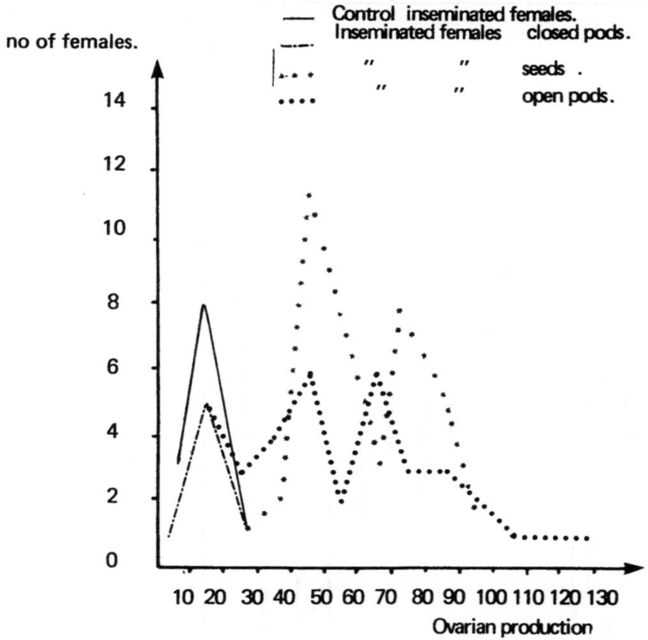

Fig. 3. Distribution of ovarian production of inseminated females in presence of pods or bean seeds.

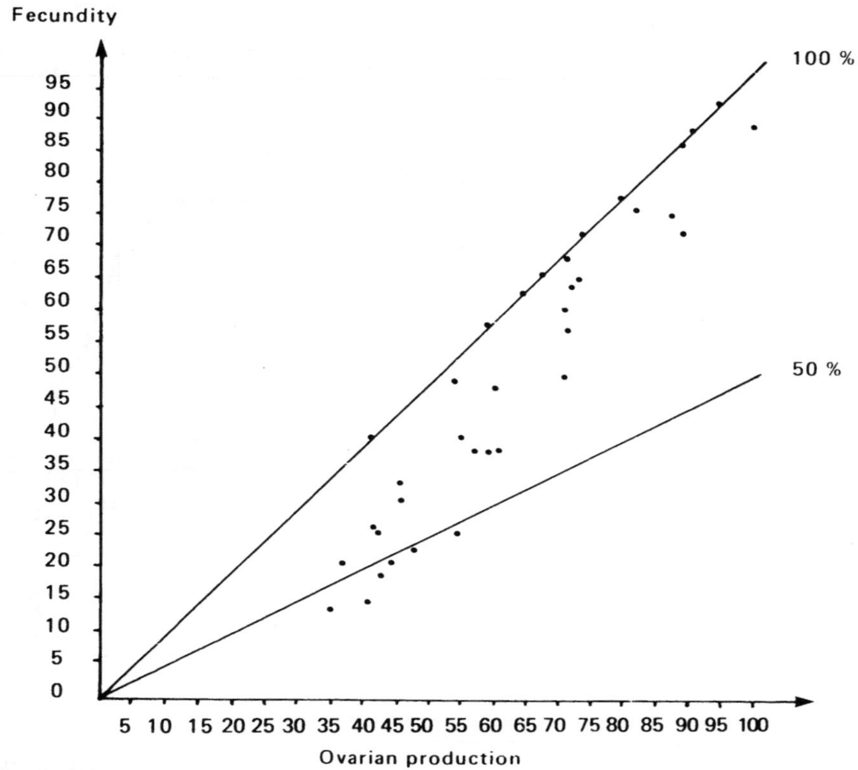

Fig. 4. Relation between fecundity and ovarian production of inseminated females.

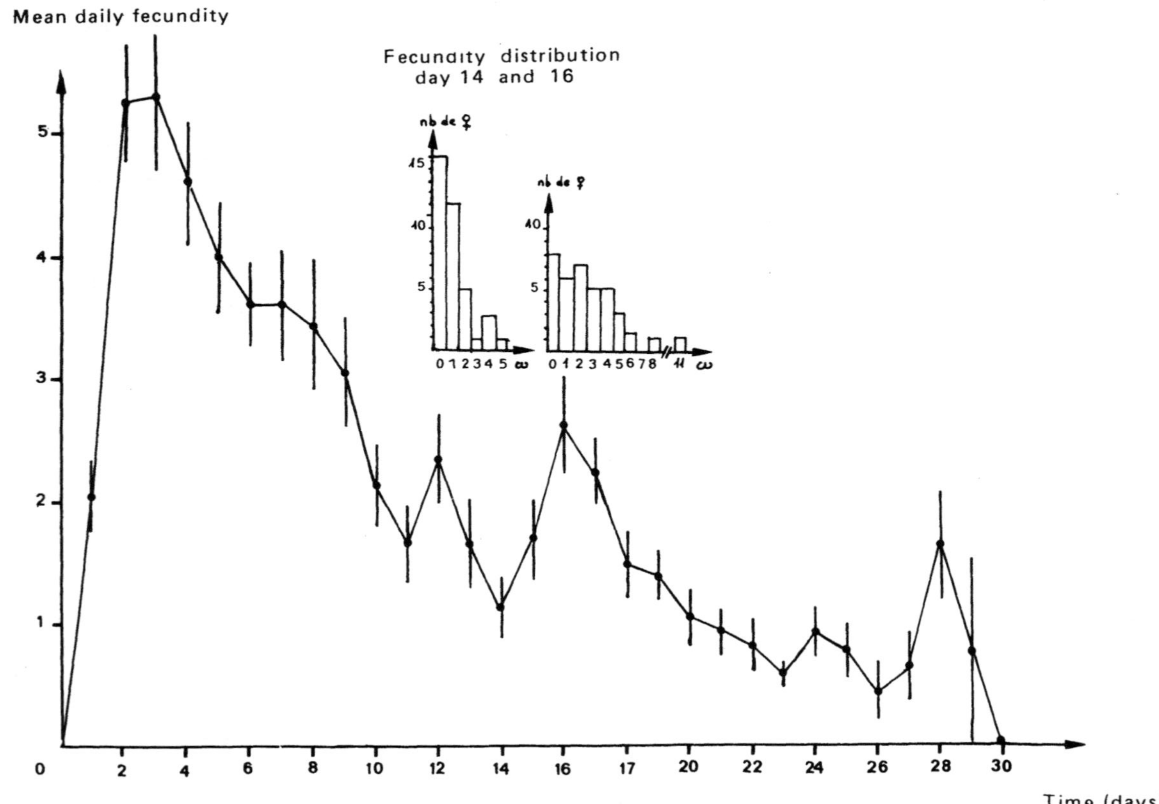

Fig. 5. Number of eggs laid by *Zabrotes subfasciatus* over time (Mean daily values + SEM).

Egg laying rythm. Under these experimental conditions, female egg-laying activity was high during the first 3 days following insemination and then gradually decreased.

However, females rarely showed a continuous egg laying activity, egg laying intervals alternated with periods of rest (Fig. 5). This egg laying pattern is roughly similar to what has been observed in other Bruchidae (Huignard, 1973; Ouedraogo, 1978).

DISCUSSION

Ripe non-dehiscent pods of *Phaseolus vulgaris* have a limited influence on the sexual behaviour and oogenesis of *Zabrotes subfasciatus*, unlike what has been observed for Acanthoscelides obtectus (Jarry, in press). Insemination alone slightly stimulates oogenesis but does not affect egg laying. The mode of action of this stimulation is still unknown for this insect; male secretions or sperm may be involved as has been shown in some Bruchidae (Huignard, 1975) and in other insects (Leopold, 1976).

On the other hand, free seeds or open pods making bean seeds more accessible facilitate sexual meetings, induce egg laying and provoke an increase in ovarian production. Contact with bean seeds therefore seems necessary to induce reproductive activity.

Under these conditions, eggs are laid directly on the food base of the larvae. This would suggest that in nature *Zabrotes subfasciatus* populations maintain themselves on wild or sub-spontaneous legumes whose pods are partially or totally dehiscent.

Although the oviposition period lasts for 17 to 30 days, females lay most their egg during the first 5 days of their life. Egg laying activity is not a continuous phenomenom. Oviposition periods alternate with intervals of rest. These interruptions in egg laying activity could explain the higher fecundities and longevities recorded in our experiments compared with previous studies. Fecundity values quoted in the litterature, for example, fluctuate between 20 and 50 with maxima not exceeding 69.

Such marked differences between our results and previous values compel us to be critical of studies which typify species in terms of average fecundities longevities and optimum developmental temperatures (Labeyrie, 1960).

Unlike *Acanthoscelides obtectus* (Labeyrie, 1962), *Zabrotes subfasciatus* cannot perforate pods in order to lay its eggs nor can it lay eggs on pods as does *Callosobruchus maculatus* (Ouedraogo, 1978). Females must therefore search for partially or totally dehiscent pods. The use of totally non-dehiscent varieties favouring a spatial separation between the reproductive and consumer stages of this insect could contribute to reducing contamination before the harvest. Moreover, cultural methods aimed at controlling wild dehiscent legumes which could act as reservoirs for this insect around farms should be studied.

These propositions should be tested in the field, on legume crops, taking into account the role of other phytophagous insects (Singh & Van Emden, 1979; Van Emden, 1980) such as the Lepidoptera which can make bean seeds accessible to bruchids by perforating pods.

REFERENCES

Aykroyd, W.R. & Doughty, J., 1964. Legumes in human nutrition. FAO Nutritional studies No. 19. FAO, Rome, Italy.

Balachowsky, A.D., 1962. Traité d'entomologie appliquée à l'agriculture, pp. 434–494. Masson, Paris, 559 pp.

Bondar, G., 1937. Notas biologicas sobre Bruchideos observados no Brasil. Arc. Inst. Biol. Veg. 3 (1): 1–44.

Bridwell, J.C., 1918. Notes on the Bruchidae and their parasites in the Hawaiian islands. Proc. Hawai Ent. Soc. 3: 465–505.

Binaghi, G., 1955. Sulla ricomparsa in Italia dello Spermaphagus subfasciatus Boh. e raffronti con un attro Bruchidae nostrano del fagiolo. Boll. Lab. Ent. Agri. 14 (seen in abstract only.)

Carvalho, R.P.L. & Rossetto, C.J., 1968. Biologia del Zabrotes subfasciatus Boh (Coleoptera Bruchidae). Revta. Bras. Ent. 13: 105–117.

Davies, J.C., 1972. A note on the occurrence of Zabrotes subfasciatus Boh., Coleoptera (Bruchidae) on legumes in Uganda. East African Agric. and fore. J.: 294–299.

Howe, R.N. & Currie, J.E., 1964. Some laboratory observations on the rates of development, mortality and oviposition of several species of Bruchidae breeding in stored pulses. Bul. Entomol. Res. 55 (Part 3): 437–477.

Huignard, J., 1973. Influence de quelques stimulations externes dues à la copulation sur la fonction reproductrice des femelles chez Acanthoscelides obtectus (Say) Coleoptera – Bruchidae. Thèse Doctorat d'Etat, Tours, 206 pp.

Huignard, J., 1975. Influence de la copulation sur la fonction reproductrice femelle chez Acanthoscelides obtectus. Ann. Sci. Nat. Zool. 1711: 1–66.

Jarry, M., in press. Comparison des fecondites et des pontes d'Acanthoscelides obtectus Say (Coleoptera-Bruchida) en presence de graines et gousses de Phaseolus vulgaris.

Labeyrie, V., 1960. Contribution à l'étude de la dynamique des populations d'insectes: influence stimulatrice de l'hôte Acrolepia assectella Z. sur la multiplication d'un hyménoptère Icheumonidae (Diadromus sp) Entomophaga – Mem. 1, 193 pp. Paris.

Labeyrie, V., 1962. Mise en evidence d'influences multiples de la plante-hôte sur la stimulation de la ponte chez Acanthoscelides obtectus Say C.R. Soc. Biol. 156 (8–9): 1473–1477.

Leopold, R.A., 1976. The role of male accessory glands in insect reproduction. Ann. Rev. Entomol. 21: 199–221.

Lepesme, P., 1944. Les coléoptres des denrées alimentaires et des produits industriels entroposés (P. Lechevalier).

Mendes Ferreira, A., 1960. Subsidios para o estudo de nua proga do jegao (Zabrotes subfasciatus Boh-Coleoptera-Bruchidae) des climas tropicais. Garcia de Orta 8 (3): 559–581.

Ouedraogo, P.A., 1978. Introduction à l'étude de quelques aspects de la biologie de Callosobruchus maculatus (coléoptre-Bruchidae) et de l'influence de facteurs externes stimulants, plante hôte et copulation, sur l'activité reproductrice de la femelle. Thèse 3eme cycle, Toulouse, 1–101 pp.

Pierre, D., 1979. Contribution à l'étude de l'activité reproductrice males chez Zabrotes subfasciatus Boh. D.E.A. 1–44 pp. Tours.

Singh, T., Kaur, I. & Saini, M., 1979. Biology of Zabrotes subfasciatus Boh (Bruchidae-coleoptera). Entom. 4 (2): 201–203.

Singh, S.R. & Van Endem, H.F., 1979. Insect pests of grain legumes A. Rev. Entomol. 24: 255–275.

Steffan, J.R., 1945. Contribution à l'étude de Zabrotes subfasciatus Boh Mem. Museum Hist. Nat. XXI 2, 55–84.
Van Emden, H.F., 1980. Insects and mites of legume crops, pp. 107–192. In: Advances in Legume Science R.J. Summerfield & A.H. Bunting, (eds.).
Zacher, F., 1930. Untersuchungen zur Morphologie und biologie der Samenkäfer (Bruchidae-Lariidae) Arb. Biol. (Anst. Reichsdnst) Berlin 18 (3): 233–284.

Authors' address:
I.B.E.A.S. (ERA CNRS 328)
Université François Rabelais
Parc de Grandmont
37200 Tours
France

TROPHIC RELATIONS AND ECOLOGICAL STATUS OF THE ADULTS OF *BRUCHUS PISORUM* L. AND ALLIED FIELD SPECIES OF BRUCHIDAE (COLEOPTERA)

H.R. PAJNI
(*Chandigarh, India*)

INTRODUCTION

Bruchus pisorum L., commonly called as Pea-bruchid, is a serious pest of peas, attacking the green pods in the field and completing its development partially in the field and partially in the harvested seeds. The species is univoltine and completes a single generation in about six months between February to August. The males and the females emerging from the infested seeds pass the unfavourable six months of the year in a condition of what may be called as 'imaginal diapause'. Such adults are reported to hide themselves under dead leaves, bark of trees, cracks in soil, splinters of fence posts etc. (Wakeland, 1934; Baranyovits, 1944; Al-Ravy & Kaddou, 1971) or to burrow in soil (Vasilev, 1939). According to some reports, the adults during this period also resort to feeding on flowers and leaves of plants other than *Pisum sativum* (Reichart, 1964). It is also a common observation that the males and the females have to feed on the pollen of pea plants in order to be able to copulate (Vasilev, 1939; Reichart, 1964; De Luca, 1966). There is, however, hardly any report on the hibernation and feeding behaviour of the bruchids during the complete annual cycle. The present communication accordingly records some of the observations made on the diapause development and the feeding activity of *Bruchus pisorum* with particular reference to their obligatory pre-copulation feeding on the pollen of peas.

The males and females of *B. pisorum* used in the present study, emerged from the sample of infested pea-seeds procured locally from July to September. A few samples procured from cold localities like Solan (height 1300 m) and Simla (height 2000 m) showed that the emergence of adults continued up to late February and the adults from this emergence were largely used for observations on the feeding of the adults on the pollen of pea plants. The trials on the selective feeding by the males or the females were conducted in the pea-fields by confining the bruchids in small muslin-bags tied around the pea-flowers.

OBSERVATIONS

The pea-fields around Chandigarh, a modern town located at the foot of the Shivalik range (latitude 30° 45′ north; longitude 76° 45′ east) bloom in the

Series Entomologica, Vol. 19: ed. by V. Labeyrie

second half of February and the formation of the pods is completed by the end of the first week of March. The males and the females are seen feeding on the pollen of pea flowers and the females ovipositing on the surface of young pods. The crop is harvested in the months of April and May, the exact time depending on the length of the winter season. The freshly harvested seeds usually contain the 2nd or 3rd stages of larvae and their presence inside the seeds is indicated by the formation of small blackish dots at the places of their entry into the seeds. The emergence of the adult weevils through the circular windows, prepared by the mature larvae, starts during the month of August (Temperature about 35°C; relative humidity high). The adult bruchids emerging from samples collected from local shops and markets, generally leave the seeds and fly away into the open where they presumably hide themselves under the dry vegetation or the bark of trees. In spite of vigorous search, no such over-wintering bruchids came to notice of the author and hence the exact hiding places of the bruchids are obscure. It was, however, observed that the emergence of adults from the infested samples of peas procured from colder places such as Solan and Simla was sufficiently delayed and continued up to the middle of February.

The males and females emerging during the months of August and September and kept in the laboratory died after a few days. However, the life-span of the adults was prolonged by about 10 days, when these were provided with a few dry leaves under which the bruchids were usually found hiding. Some of these bruchids were separately provided the flowers of such plants as cotton (*Gossypium hirsutum*), lady finger (*Abelmoschus esculentus*), and shoe flower (*Hibiscus rosasinensis*). The bruchids were found to feed on the pollen grains of these flowers and were able to maintain themselves for much longer periods as compared to those which were not provided with these flowers and were therefore short lived. The development of gonads was studied in the laboratory-fed males and females between October and February at fortnightly intervals. Whereas the testes of even the freshly emerged males were found to be packed with bundles of mature spermatozoa, the ovaries showed hardly any development and there was no trace of the formation of ova throughout this period. However, when the females were later on fed on the pollen of pea, there took place a rapid development of ovaries which became mature after a few days feeding. It is thus evident that the ovaries remain in a state of dormancy throughout the six month period of inactivity while the testes show full development from the very beginning. It was further noticed that the feeding on pollen of peas by both the sexes is essential for inducing copulation in them. This has been verified by dividing the bruchids into the following four lots, depending on the conditions of feeding of adults.

1. Both males and females were fed on pea pollen.
2. Only males were allowed to feed on pea pollen.
3. Only females were fed on pea pollen.
4. Both males and females were not fed on pea pollen.

The males and females from all the four lots were paired after every two days for observing the occurrence of copulation. It was discovered that the copulation took place among the individuals of the first lot and the sexes copulated after feeding on pollen for 8 days. No copulation occurred in the remaining three lots, even after allowing feeding for 18 days to the individuals of the concerned sex in the lot Nos. 2 and 3. Apart from visual observations, the occurrence of copulation or otherwise was also confirmed by examining the spermathecae of the females which showed the presence of spermatozoa in the copulated ones. This clearly shows that feeding on the pollen of peas by both the sexes is necessary for copulation and that non-feeding on pea pollen or selective feeding by one of the sexes fails to bring about copulation. The males and females copulate during day time and several pairs confined in a tube in the late evening were found to copulate early next morning. Likewise, the females also lay their eggs during day time. The feeding adults move from one plant to another by taking slow and short flights, but the adults are able to fly at a great speed when shifting from one field to another field at a distance of about 20 m. The flight activity of bruchids becomes particularly pronounced during afternoons as compared to morning and evening when they remain more or less settled.

DISCUSSION

The peas are sown during different months in different areas under varying climatic conditions. Accordingly, the active part of the life span of *Bruchus pisorum* coincides with the availability of its host plants and the dormant phase of its life is passed in winter or summer months. This bruchid species is known to be univoltine, passing only a single generation in the course of a year and thus belongs to the category of 'field forms' (Luk'yanovich & Ter Minassian, 1957; Southgate, 1979). The adults emerging from the harvested seeds enter a period of inactivity during which they generally hide themselves under sheltered places like dead vegetation, bark of trees, crevices in the wood or soil etc. (Whithead, 1930; Larson, 1931; Brindley, 1933, 39; Wakeland, 1934; Baranyovits, 1944; Hinman *et al.*, 1949; Al-Ravy & Kaddou, 1971). These adults thus overwinter in a dormant state, stop reproduction and show the phenomenon of 'imaginal diapause'. According to Vasilev (1939), such adults actually burrow in the soil to pass a period of true hibernation. The author has failed to observe any such overwintering adults and therefore nothing definite can be said about their exact hiding places. However, it has been observed that the emergence of the adults is delayed in samples of infested peas procured from colder regions in which some of the adult bruchids are found reoccupying the cavities of the seeds, after emergence.

Observations on the behaviour of the diapausing adults under captivity in the laboratory indicate that the early death of the adults released into glass containers definitely points to the necessity of seeking shelters by the bruchids in order to pass the adverse season. In fact, the prolongation of the life-spans of these adults, when they are provided with some sheltering material consisting of dry leaves and grasses, affords an experimental support to this aspect of their behaviour. It has also been observed that the adults with the sheltering material were also short-lived and survived only when flowers of different available plants such as *Gossypium hirsutum* (Cotton) *Abelmoschus esculentus* (Lady finger) and *Hibiscus rosasinensis* (Shoe flower) were added as food. The bruchids have been observed to actually feed on the pollen of these flowers which is further evidenced by the examination of their gut contents. The feeding of the adults on the pollen of eleven different flowers in the laboratory has also been noticed by Vasilev (1939). Such a feeding by the adults on flowers of plants other than the host plant *Pisum sativum* has not been noticed by any worker in nature, although Reichart (1964) reports the feeding of the adults on 'weeds' before their shifting to pea fields. The possibility of the feeding of the bruchids on the pollen of different flowers in the laboratory, as observed by the author & Vasilev (1939), and its definite role in the prolongation of the life-span in the former case, raises two important questions. The experimental proof of the important role of the feeding of the bruchids during their inactive phase points to the inevitable conclusion that the adults hiding under different shelters might be resorting to occasional feeding on different flowers during the dark hours of the night, in order to avoid death by starvation. Moreover, the acceptance of the pollens of other flowers by the bruchids shows the distinct possibility of the emergence of such plants as secondary host plants of *Bruchus pisorum,* at least for the purpose of tiding over the non-vegetative period of their preferred host plant.

It has been noticed that the male bruchids are sexually mature at the time of their emergence and their testes contain mature spermatozoa, while the females possess only undeveloped ovaries without any indication of the formation of ova. Moreover, the feeding of the adults on the pollen of non-preferred flowers neither results in the maturation of the ovaries nor in the induction of copulation in the two sexes. A further probe into the relationship between feeding on pea pollen and copulation has revealed that feeding on pea pollen by both the sexes is a prerequisite for copulation. According to Vasilev (1939), the females fed in the laboratory on the pollens of 11 different flowers other than that of peas did not show the development of their ovaries till they had fed on the pollen of peas for several days. On the other hand, Reichart (1964) and De Luca (1966), have stated that feeding on the pea flowers is essential for the development of gonads in both the sexes. The present investigations have accordingly brought to light an interesting situation in which the feeding of *Bruchus pisorum* on the pollen of the host plants completes the maturation of ovaries, but at the same time, fulfils in the two sexes, certain physiological conditions essential for copulation.

Another allied monophagous species viz., *Bruchus lentis* Froel, feeds on the seeds of *Lens culinaris* (*Masur*) and shows a more or less similar univoltine cycle. However, the emergence of the adult bruchids in this case starts soon after the harvesting of the pulse in the month of April/May. The hiding places of the adults have not been traced, but the same have been observed to visit in large number the fields of the host plant to feed on the pollen of its flowers.

Some observations have also been made in this laboratory on the biology of a few more field species such as *Bruchidius angustifrons* Schilsky, *B. pygomaculatus* Arora and *B. mimosae* Arora, which are associated with the pods of wildplants and are thus not of much economic importance. *B. angustifrons* attacks the pods of *Sesbania seban L.* which is only grown on a limited scale usually for the purpose of green manuring or as a fodder plant for goats etc. This species undergoes as many as three or four generations in the course of a year depending on the number of bloomings in the host plants. The adult bruchids emerging from different generations actively feed on the pollen of *S. sesban* but the feeding does not seem to be essential for copulation in this species. *B. pygomaculatus* is normally a univoltine species and is associated with the pods of *Albizzia lebbek* Benth, *Albizzia procera* Benth, *Acacia modesta* Wall and *Acacia catechu* Wild., but sometimes passes 2 successive generations on the pods of the first mentioned host plant. Feeding on the flowers of the host plants has been noticed but appears to be unnecessary for copulation. *B. mimosae* is again a univoltine species, attacking the green pods of *Mimosa rubicaulis* Lamk. Feeding on the pollen has been noticed but the same is not essential for copulation. It is thus evident that the field forms of *bruchids* differ in the number of generations passed and in the influence of host plant feeding on the initiation of copulation. This variation may be due to the degree of host specificity exhibited by different species, but there may be other physiological and ecological factors involved in this process.

REFERENCES

Al-Ravy, M. & Kaddou, I.K., 1971. The pea-weevil *Bruchus pisorum* L. (Coleoptera: Bruchidae) infesting *Vicia faba* L. in Iraq. Acta Entomol. Bohemsolov. 68 (6): 365–371.

Baranyovits, F., 1944. Bionomics and control of pea weevil. Year book Phytosanit Serv., Budapest 2–4: 268–276.

Brindley, T.A., 1933. Some notes on the biology of pea weevil, *B. pisorum* at Moscow, Idaho. J. Econ. Ent., Geneva, N.Y., 26: 1058–1062.

Brindley, T.A., 1939. Biology and control of *B. pisorum* in the Palouse area of Idaho and Washington. J. Econ. Ent. 32: 21–24.

De Luca, Y., 1966. Alimentation imaginale des Bruchides (Col.). Parasitica 22 (1): 26–54.

Hinman, F.G., Brindley, T.A. & Schopp, R., 1949. Hibernation of the pea-weevils. J. Econ. Ent. 42 (5): 746–753.

Larson, A.O., 1931. Pea-weevil control in Willamette Valley. Sta. Circ. Oregon. Argric. Expt. st. 99.

Luk'yanovich, F.K., & Ter Minassian, M.E., 1971. Fauna of the U.S.S.R.: Coleoptera: Seed beetles. Akad. Nauk. U.S.S.R. Zool. Inst. Fauna U.S.S.R. 24.

Reichart, G., 1964. Comprehensive study on *B. pisorum* and its control. Kaiseri. Kozl. Budapest (A), 51: 149.

Southgate, B.J., 1979. Biology of the Bruchidae. Ann. Rev. Entomol. 24: 449–73.

Vasilev, I.V., 1939. The results of investigations upon the pea-weevil. Plant B. Prot. for the year 1939. Lenin Acad. Sci. 8: 184.

Wakeland, C.C., 1934. Flight studies on *B. pisorum.* J. Econ. Ent. 27 (2): 534–542.

Whithead, F.E. 1930. Pea weevil problem. J. Econ. Ent. 23: 398.

Author's address:
Zoology Department
Panjab University
Chandigarh-160014
India

EVOLUTION OF SPATIAL PATTERN OF ATTACKS BY *ACANTHOSCELIDES OBTECTUS* SAY (COLEOPTERA: BRUCHIDAE) OF *PHASEOLUS VULGARIS* L. PODS IN SOUTH WEST FRANCE

M. JARRY
(Tours, France)

INTRODUCTION

When listing the enemies of pulses, Bruchidae are generally registered at the end of the list under the term of 'storage pests'. To consider a very recent example, Singh & Allen (1980) include *Callosobruchus maculatus* and *Callosobruchus chinensis* in this category when referring to the damage caused to the cowpea (*Vigna unguiculata*) while Larson & Fisher had already described in 1938 the oviposition of *Callosobruchus maculatus in* the fields on *Vigna* pods.

Regarding the bean weevil *Acanthoscelides obtectus* a rapid survey by an automatised system of the works published in the past ten years shows that most of them have been directed to the methods of control in stocks.

Though self-infestation may occur in stocks, where a succession of crops can be accumulated and mixed without elementary precautions, infestation generally starts in the fields (where inseminated female- lay their eggs in ripe and dry pods (Labeyrie & Maison, 1954a). In South West France, infestation is caused by adults which fly away from storage areas (Arnoux *et al.*, 1958).[1]

The present study was conducted to know the distribution in time and space of weevil attack in *Phaseolus vulgaris* crops, and the possible influence of climatic conditions prevailing in the period under study on this phenomenon.

The question of spatial relationships between insects and plants though well documented in ecological works, has not often been studied closely in nature, and as was rightly emphasized by Legay and Chessel (1977) 'Few methodological approaches have been proposed to study the relations between an animal population and the vegetation to the presence-absence qualitative links on a plant or part of it, or on the host for parasity'. The information is reduced to a statistical distribution which thus overlooks the role of space.

[1] Infestation may occur from sown seeds as the latter can contain almost fully-grown weevils (Zachariae, 1958, 1959). This type of infestation seems to be highly questionable because those seeds rapidly get soaked, owing to be tegument injuries increasing the absorbing capacity and thus favouring considerably the development of bacterie and bacterial and fungal development (Gain, 1897).

This approach is illustrated by a number of articles in the first volume of *'Statistical ecology'* published in 1971, yet called *'Spacial patterns and statistical distributions'* (Bliss, 1971; Taylor, 1971).

Now any attempt to develop ecological methods for crop protection should examine to what extent the level of attacks can be reduced by utilizing landscope heterogenity, thus avoiding all systematic chemical interventions, or at least reducing them to punctual treatments perfectly localized in space.

Therefore, in order to take as much as possible into account space and time dimensions, graphic repose sensation were often used, linked to simple statistical tests (contingency table, mean and frequency comparison tests).

The results will be discussed taking into account both field observations and what we know on the biology of *Acanthoscelides obtectus,* and compared with the distribution of pea-weevil *Bruchus pisorum* attacks (Larson *et al.*, 1933) the only available instance for Bruchidae, to our knowledge.

MATERIALS AND METHODS

Field work (pickings and observations) was conducted in Aire sur Adour (latitude 43° north) in the South West France. This is an area with a bocage characteristic which has been progressively removed by an ill planned regrouping of lands, was among the more important in France for the production of beans. The almost permanent presence in this area of bean weevils, which were apparently first observed in France in 1873 in the department of Pyrenees Orientales (Girard, 1879) has been one of the main factors which has caused the cultivation of beans to decrease.

Nowadays beans are being cultivated only for domestic consumption, which is still sufficient to ensure the maintenance of weevils.

The climatic conditions were rather exceptional for the period under study (September, mid-October 1978) with hardly any rain and very sunny afternoons after dispersion of morning mists; Though the nights were cool (mean minima: 9 °C with 4 °C extrema) temperatures above 25 °C with 35 °C maxima, were often recorded in the afternoon (Fig. 3).[2]

The lots, of about 125 m^2 were made up of 4 rooms of 16 seed holes containing 5 seeds each of a variety of climbing bean – 'French bean' – whose pods are, according to Evans's classification (1975) of an intermediate type between the very dry and dehiscent ones of the varieties used for the production of seeds and these, very little stringy, of varieties cultivated for the consumption of green pods.

The rows were 1.50 m apart, and so were the seed holes in each row. These lots were isolated in the middle of several acres of corn and located along a hedge.

Two harvesting techniques were used:

LOT A: Every other day, all the pods reaching maturity (that is yellow and quite dry) were marked and harvested five days later. The heights of the

[2] These climatic data were recorded on the lot under a make-shift shelter, and hence are not standardized. They provide, anyway, a rather good picture of the climatic conditions that prevailed during the period under study.

points where the pedoncules of pods were attacked to the stalks, were recorded. Thirteen pickings were thus made from 12th September till 6th October 1978.

LOT B: All the pods were harvested at the same time in 20 cm strata simultaneously to the last picking in lot A on 27 plants chosen at random.

All the crops were taken to the laboratory, where the emergences of adults for each pod were counted.

RESULTS

The attack was evaluated by the frequency of pods giving one emergence or more. This frequency can be calculated globally, either per plant, or harvesting date, etc. The mean number of weevils per attacked pod provides complementary information. This value can actually vary in the same attack.

The data yielded by Lot A, being more complete, will first be analyzed in detail, then the results will be compared with those yielded by Lot B, whose pods were longer exposed to weevil attacks.

Fig. 1 gives a simplified record of the pickings and attacks in which the information is reduced to statistical variables of the presence-absence type (picking or not, attack or not). Yet it can be noticed that the evolution of pod ripening was not similar in the various plants. Row a appeared to be late. Yet on the whole, the frequency of attacked plants remained nearly constant throughout the pickings ($\chi^2 = 10.39$ for χ^2 th. = 19.67 at 5%).

This led us to analyse the relationships between the time of pod ripening and the rate of attack for each plant.

The analysis of correspondences as a method for processing using a table (Esteve, 1978) enables us to classify the plants according to the first factor given by analysis (which is obviously to be taken as the time factor).[3] This classification showed the existence of early or late ripening plants. There was no break in the transition from one to the other of these extreme types, but there was no connection between this phenomenon and the global level of attack (in the 13 pickings).

This lack of relationships at the plant level between the rate of attack and the ripening process of pods led us to carry on the analysis in a more general way, taking into account for the analysis in time, all plants together, and for an analysis in space, all the crops of the lot together.

Evolution of attacks in Lot A

When following the curve of evolution of attacks (Fig. 2) a slight decrease could be inferred for the first four crops, followed by an increase for the subsequent ones. In effect a χ^2 test showed that the distribution of the attacked pods varied with the crops ($\chi^2 = 34.57$ for a χ^2 th = 19.67 at 5%). But this variation came mainly from two crops (No. 3 and 11), the biggest one

[3] The calculation were made by D. Chessel in the Biometry Laboratory in Lyon.

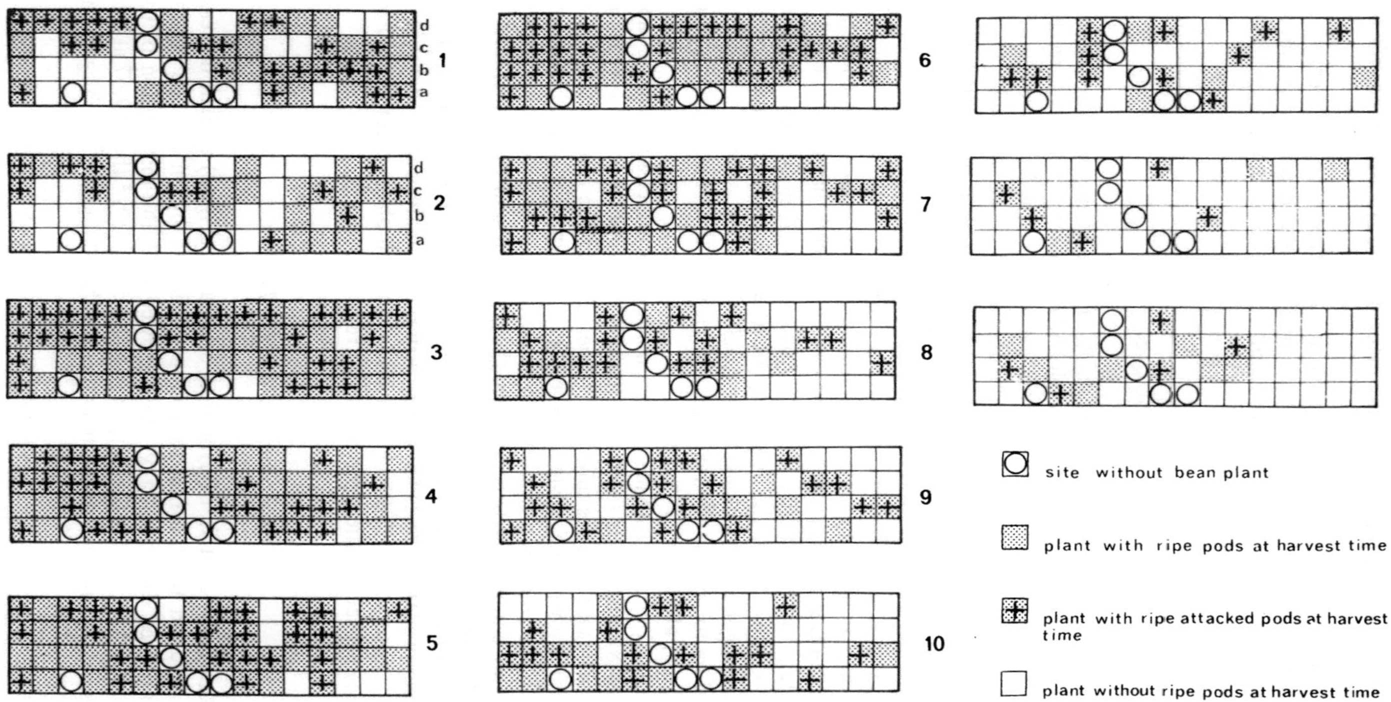

Fig. 1. Evolution in space and time of the presence of ripe pods of *Phaseolus vulgaris* and attacks of *Acanthoscelides obtectus* during 13 consecutive crops in Lot A.

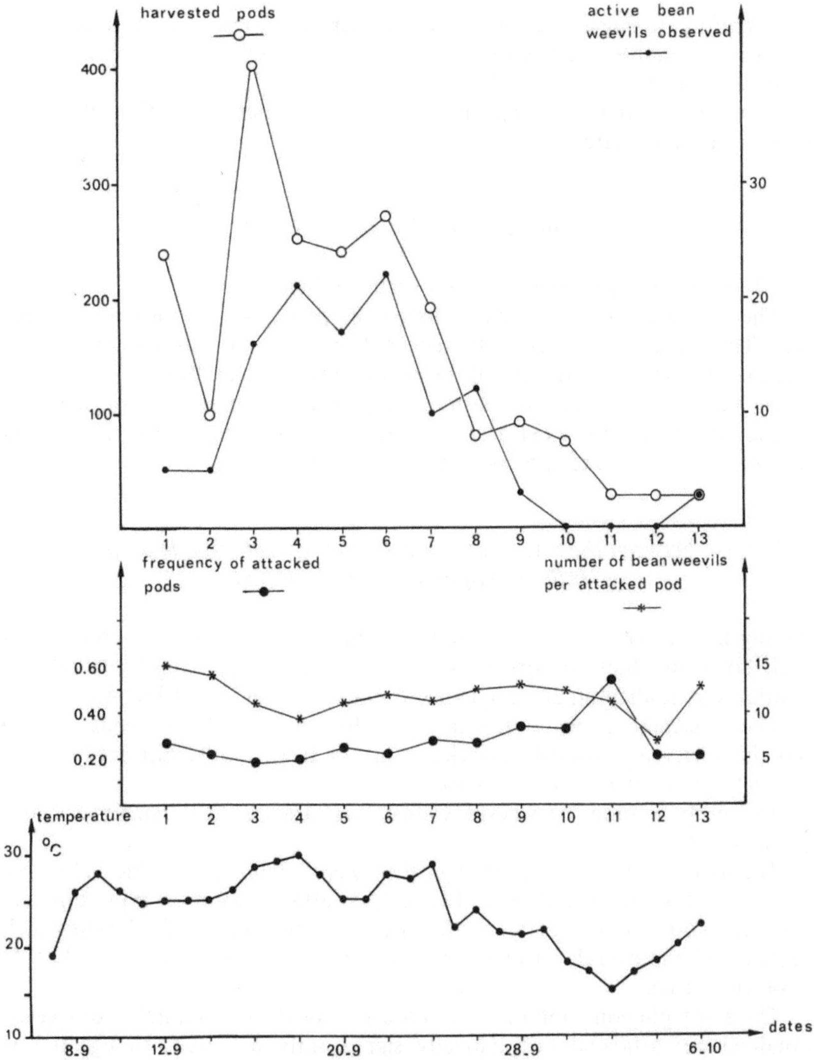

Fig. 2. Evolution of climatic conditions, of the number of harvested ripe pods of *Phaseolus vulgaris* and of attacks by *Acanthoscelides obtectus* on Lot A.

and one of the smaller ones. If we leave aside those two values, the χ^2 is no longer significant ($\chi^2 = 14.50$ for a χ^2 th $= 21.51$ at 5%).

As the ovipositional possibilities of females were limited to a few hot hours in the afternoon by the periodicity of favourable conditions, we made a rough estimate of the mean temperature of these hotter hours (Fig. 2). The period under consideration was generally highly favourable to the activity of *Acanthoscelides obtectus*, and even though the mean temperature was below 20 °C in the last days, that did not prevent the attacks.

Therefore, taking into account the considerable variation in the importance of crops, a relative stability of attacks during the period under consideration could be inferred.

The mean number of emergence, which was higher in the first two crops hardly ever varied later.

Distribution of attacks in Lot A

Fig. 3 shows a tendency that was noticeable in Fig. 1.

There seemed to be fewer attacks on the first row, and more on the last one. The value of χ^2 testing the equidistribution of the attacks among the 4 rows was highly significant (40.00 for a χ^2 th = 7.81 at 5%).

In addition the first row eventually yielded fewer pods, though it was evident there did not exist at the plant level any conspicuous relationship between attack and the total pod yield per plant.

Relation between the rate of attack and the position of pods on the plant: Influence of exposure time of pods

Taking into account the above results, and in order to draw a comparison with Lot B in which the first pods were left for a longer length of time this relation was studied on 26 sample plants chosen at random in Lot A.

There seemed to be fewer attacks on the lower and higher pods (Fig. 4) but this difference, though remarkable in Lot B was not significant in Lot A (χ^2 = 9.82 for χ^2 th = 14.06 at 5%).

The mean number of weevils yielded per attacked pod remained nearly constant.

The lower pods in Lot B, though they were left longest in the field were not more attacked than those in Lot A. Most attacks occurred approximately between 0.40 m and 1 m. It was the difference observed at this height which determined the overall difference in the attack frequencies (0.25 for Lot A; 0.34 for Lot B).

The same phenomenon could be observed for the mean number of weevils obtained per attacked pod, globally significantly different, though it was higher for the 2 lots (12 – Lot A –, and 14 – Lot B – weevils per attacked pod).

DISCUSSION

First the remarkable capacity of *Acanthoscelides obtectus* adults to find bean plants is to be noticed. In effect our lots were isolated in the middle of corn fields, in a 1/10 000 acreage ratio, and over 500 m away from possible sources of infestation. This did not prevent 57 plants out of 58 being colonized by adults.

In Lot A, the pods were harvested five days after ripening. Thus a few

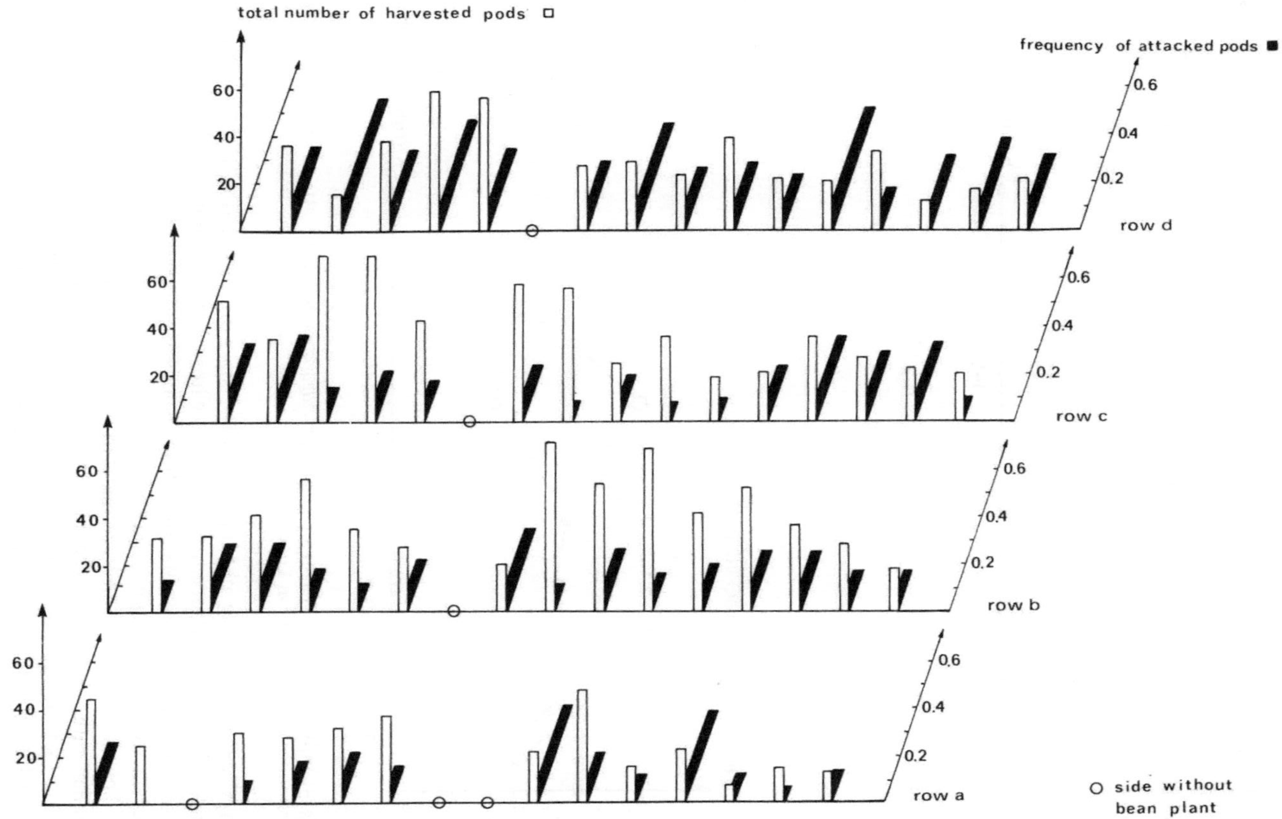

Fig. 3. Spatial pattern of total pod production per *Phaseolus vulgaris* plant and distribution of the attack frequency of *Acanthoscelides obtectus* in Lot A during 1978 season.

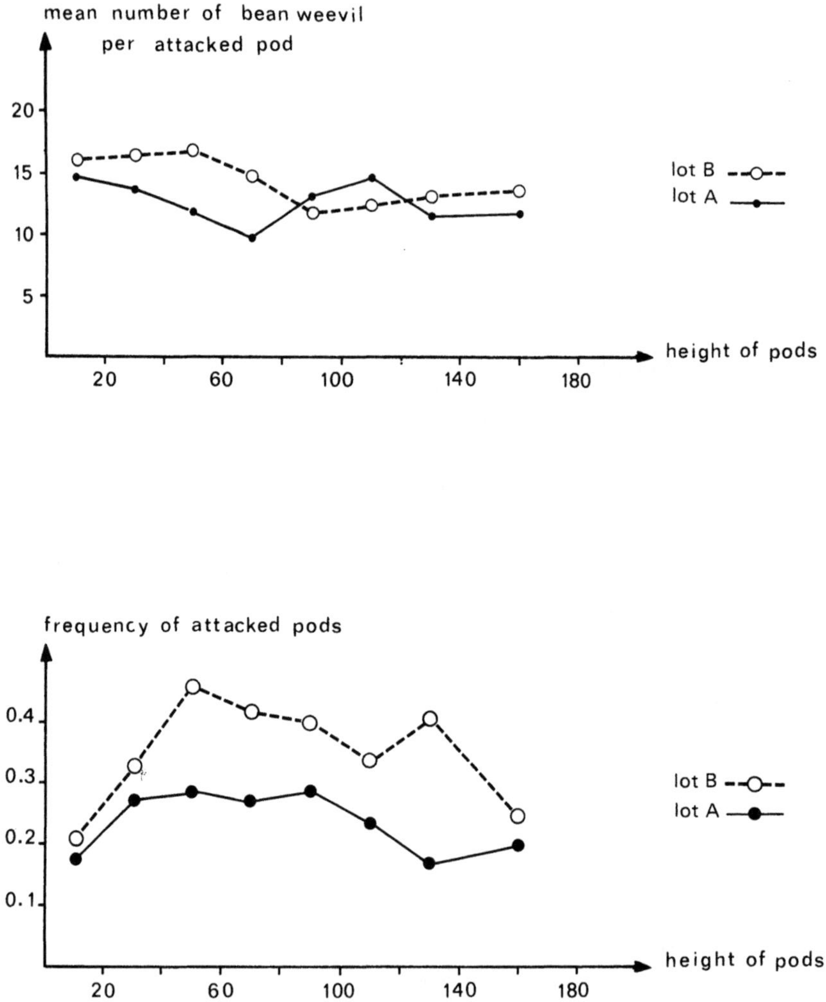

Fig. 4. Influence of the height of *Phaseolus vulgaris* pods on the attack frequency of *Acanthoscelides obtectus* in Lots A and B.

days' exposure at the phenological stage allowing oviposition (let us recall that 95% egg layings occur in dry pods whose water content has dropped to about 4% (Labeyrie & Maison, 1954a) pods were attacked and a mean 12 weevils were counted per attacked pod).

This attack was observed as early as the first crop with an almost similar rate. It implies the insects must be present on *Phaseolus vulgaris* plants before the pods are ripe, or if they are near the lots, they respond immediately after the ripening of pods by moving towards the bean plants, which bear out the observations that have been made of the periods of flight away from stores (May–June 1952 and 1953 according to Labeyrie (1953),

July 1978 and 1979 in Aire-sur-Adour for a type of granary rather cold and damp and poorly sunlit, which accounts enough for that difference).

In addition, it implies an immediate responsiveness of insects as soon as the sensitive phenological stage has been reached. In effect all the females collected in the lots had been inseminated. This could be verified in 1978 with a similar plant distribution; all the females collected in the early period of attack and subsequently dissected contained one or several spermatophores.

The harvesting methods used in Lot A should concentrate the adults on the plants that still bear pods. Under such conditions, the attack could be expected to augment considerably in the last crops, owing to the insects moving from the already harvested plants to those still bearing seeds. There the later plants should be the more infested as well.

But we have seen that it was by no means the case since the rate of attack was remarkably constant in space and time. Two complementary hypotheses can be formulated to explain this phenomenon:

– Once the insects have colonized the plants they do not move much in the lot. Our experiment of individual marking bore out that hypothesis (Jarry, 1980).

– The reproductive activity of females decreased during the period under study, which was very likely to occur, consequently as a result of the evolution of climatic conditions, females ageing and possibly dying and for the number of active females decreasing, as is suggested by Fig. 2.

In any case, a high insect concentration was never observed on the same plant, even after the number of those bearing pods had become very limited, at the end of the season.

Therefore, the cropping technique did not seem to alter considerably the results. This is borne out by comparing with the results yielded by Lot B.

Though on the whole the attack rate was higher, while remaining of the same order, it was remarkable that the lower pods were not more attacked, however long they were exposed: For 5 days, or 4 to 5 times as long. Now it takes a longer time for these pods to dry after the morning dew, and they are more often affected by a number of fungi.

The time of exposure played a significant role only for the pods located between 0.40 m and about 1 m above the ground. This might be related to the fact that they did not suffer from the negative effects of a prolonged dew. Hase (1951) has noted a similar distribution in Germany with the most frequently attacked pods being below 1 m above ground.

Finally, the only disturbance of spatial distribution of attacks that could be observed on row A seemed to be due to the proximity of the hedge. Now, this row remained in the shade longer than the others. The lower rate of attack could result from microclimatic factors, whose importance on the behaviour of *Acanthoscelides obtectus* had already been considered (Labeyrie & Maison, 1954b).

Therefore, these observations can be summed up by emphasizing that the distribution of attacks in the lots appeared to be very homogenous, and their importance was relatively constant during the period under study.

This type of distribution has not often been described. Aggregativeness is presented as a rule in the field of insect-plant relationships (Taylor, 1971) and the reference statistical tool is often a discrete aggregative distribution (Patil & Joshi, 1968). In fact, as in our case, in which the observed distribution was probably related to the acreage of the lots under study, the problem of scale often has to be faced when characterising a type of distribution.

However rare such result may be or not, it seems to be rather different from what is known in another Bruchidae, *Bruchus pisorum,* whose distribution of attacks is very uneven (Larson *et al.*, 1933). The borders close to the hedges where the adults can feed are very much attacked, as well as the parts of the field near the sources of infestation (such as, for instance, the trees where the adults winter under the bars.) Yet it is advisable to be cautious when making that comparison, as that study was conducted in the United States on lots covering sweveral acres, while the lots in Aire-sur-Adour hardly cover more than a hundred square metres.

Several lessons can eventually be drawn from this study regarding methods for the protection of *Phaseolus vulgaris* crops

– There is little hope of escaping the attacks of *Acanthoscelides obtectus* by moving the crop fields away from the grain storage areas or by isolating them among other crops as that insect is remarkably capable of finding its larval host-plant. Labeyrie & Maison (1954b) had noted a similar phenomenon when observing that lots established in clearings in the middle of corn fields were attacked just the same by bean weevils.

– In South West France, The period favourable to attacks can cover several months, which reduces the possibilities for avoiding infestation by late sowing (and then, late crops). A few hot hours in the afternoon are sufficient to induce the activity of insects. Very late sowing might be efficient but then the ripening of pods would be more difficult.

– The bean weevil seems to be very sensitive to any alteration of the micro-climate at the level of the bean plant. These results are to be compared to those obtained by Labeyrie & Maison (1957), who showed that the rate of attack can be considerably reduced by cultivating beans and corn as comparison crops, as the corn canopy reduces and filters the solar radiations that reach the bean plants. All this points to one line of investigation which should be pursued further in order to achieve an ecological protection of *Phaseolus vulgaris* crops against the attacks of *Acanthoscelides obtectus.*

ACKNOWLEDGEMENT

I would like to thank heartily Pierre Labeyrie, the manager of the estate of Pantagnan, where I could, thanks to his kind help, complete this study.

REFERENCES

Arnoux, J., Labeyrie, V. & Maison, P., 1958. Influence de la température et de l'insolation sur la migration des bruches du haricot (*Acanthoscelides obtectus* say) dans la nature. C.R. Acad. Sci. 247: 2343–2345.

Bliss, C.I., 1971. The aggregation of species within spatial units, pp. 311–336. In: Statistical Ecology, Vol. 1: Spacial patterns and statistical distribution, Patil, G.P., Pielou, E.C. & Watters, W.E., eds.). Pennsylvania State University Press.

Esteve, J., 1978. Les méthodes d'ordination: éléments pour une discussion. In: Biométrie et Ecologie Legay, J.M. & Tomassone, R., eds.).
Evans, A., M.A., 1975. Research on the evolution and genetic improvement of grain legumes. Dept. of Amm. Biol. Univ. Cambridge. Mem. No. 46: 5–11.
Gain, E., 1897. Sur la germiantion des légumineuses habitées par les bruches. C.R. Acad. Sc. 125: 195–197.
Girard, M., 1879. Note sur les bruches et en particulier sur la bruche du haricot. J. Soc. Horticult. Fr. 3: 95–99.
Hase, A. 1951. Die Einbürgerung des Speisebohnenkäfers als Freilandschädling in Deutschland. Trans. IX Int. Conf. Ent. 1: 666–667.
Jarry, M., 1980. Influence de la fixation par la plante-hôte larvaire, *Phaseolus vulgaris*, sur le comportement de la bruche du haricot, *Acanthoscelides obtectus* Say (Col., Bruchidae). Bull. Int. S.F.E.C.A., in press.
Labeyrie, V., 1953. Observations préliminaires sur le comportement de la Bruche du haricot *Acanthoscelides obtectus* Say, dans la nature. A.F.A.S. Cong. Lexembourg, 473–475.
Labeyrie, V. & Maison, P., 1954a. Sur les relations entre la ponte d'*Acanthoscelides obtectus* Say dans la nature et les stades phénologiques de *Phaseolus vulgaris* L. C.R. Acad. Sci. Paris 238: 1920–1922.
Labeyrie, V. & Maison, P., 1954b. De l'influence du microclimat sur la ponte de la bruche du haricot (*Acanthoscelides obtectus* Say) dans la nature. C.R. Acad. Agric., 40: 733–736.
Labeyrie, V. & Maison, P., 1957. Techniques culturales concernant la préservation des haricots contre la bruche (*Acanthoscelides obtectus* Say) Rev. Zool. Agric., 10: 1–8.
Larson, A.O., Brinfley, T.A. & Hinman, F.H., 1933. The local dispersion of the pea weevil, J. Econ. Eng., 26: 1063–1067.
Larson, A.O. & Fisher C.K., 1938. The bean weevil and the southern cowpea weevil in California. U.S.D.A. Tech. Bull. 593, 71 pp.
Legay, J.M. and Chessel, D., 1977. Description et analyse de la répartition des insectes dans une population végétale. Cas du doryphore sur pommes de terre. Bull. Ecol. 8 (1): 23–24.
Patil, F.P. & Joshi, S.W., 1968. A dictionary and bibliography of discrete distributions. Oliver – Boyd, Edinburgh, 267 pp.
Singh, S.R. & Allen, D.J., 1980. Pests diseases, resistance, and protection in cowpeas. In: Advances in Legume Science (Kew 1978) Summerfield & Butting Ed.: 419–443.
Taylor, L.R., 1971. Aggregation as a species characteristic. In: Statistical Ecology vol. I: Spatial patterns and statistical distribution, fd by Patil, G.P., Pielou, E.C. & Watters, W.E., Pennsylvania State University Press: 357–377.
Zachariae G., 1958. Das Verhalten des Speisebohnenkäfer *Acanthoscelides obtectus* Say (Coleoptera: Bruchidae) in Freien in Nord-Deutschland. Z. angew. entomol. 43: 345–365.
Zachariae, G., 1959. Kann sich der Speisebohnenkäfer *Acanthoscelides obtectus* Say (Coleoptera: Bruchidae) als Freidanschädling in Norddeutschland einbürgen? Z. angew. Entomol. 45: 55–62.

Author's address:
I.B.E.A.S. (ERA, CNRS 328)
Université F. Rabelais
Parc de Grandmont
37200 Tours
France

POLYMORPHISM AND PHASE DIMORPHISM IN *CALLOSOBRUCHUS*

S. UTIDA
(Kyoto, Japan)

INTRODUCTION

Genus *Callosobruchus* is fairly large in number of species including several serious pests of the stored beans, *C. analis*, *C. chinensis*, *C. maculatus*, *C. phaseoli*, etc. It originates in the Old World, but several of the pest species are now becoming cosmopolitan. It is middle in its body size. Development is fast and is not interrupted by diapause. Thus, several generations are usually repeated in a year. Larva is polyphagous and feeds on not only the seeds on growing leguminous plant but also its tough seeds after harvest. By the aids of these characters of food habit, the life history of multi-generations in field or storage may be capable. The author believes that the cross infestation and the phase dimorphism observed in this genus have close relation with these ecological characteristics.

POLYMORPHISM, EXCEPTING PHASE DIMORPHISM

Observations on the polymorphism have been few in the genus *Callosobruchus*, excepting the phase dimorphism in *C. maculatus.*

Possibility of the existence of cytological polymorphism was reported in *C. chinensis* and *C. maculatus* (Takenouchi, 1971, 1971a). In both species, chromosome number in females was determined as 20, while that in males studied by different authors using the materials of different geographical origin was different, 19 or 20.

In the rearing populations of both *C. chinensis* and *C. maculatus*, mutants in the body colour were observed. The black coloured adult mutant of *C. chinensis* is visually different from the wild, which is brownish. Fecundity and fertility are low in the mutant. The cross experiments showed that the mutant behaves as a single recessive autosomal one (Kasiwagi & Utida, 1972). In *C. maculatus*, three body colour mutants, red, black and white were found and these constitute a multiple allelomorph system to the colour of wild (tan). The order of dominance is red, black, white and tan (Breitenbecher, 1921).

In the rearing population of *C. chinensis*, Nakamura (1966) found the two adult types in their behaviour, the sluggish and the normal, both of which

Series Entomologica, Vol. 19: ed. by V. Labeyrie

differ not only in their reaction to light but in their fecundity. Percentage of the emergences of both types seems to relate to the rearing temperature, the larval crowding and the types of parent. The present author also found the similar variation in their general activity in the rearing populations of both *C. chinensis* and *C. maculatus* (Utida, unpublished). The percentage of emergence of both types differs from geographical origin.

In *C. analis*, two forms which like to the phase dimorphism observed in *C. maculatus* were recorded briefly (Arora & Pajni, 1959).

PHASE DIMORPHISM IN *CALLOSOBRUCHUS MACULATUS*

Differences between two forms

In *C. maculatus*, there are two distinct adult forms, which differ not only in their morphology, but in physiology, behaviour, etc. Based upon their behaviour, these were called the flightless form and the flight form by the author (Utida, 1954), and the normal form and the active form by Caswell (1960).

The morphological differences between the two forms were given at length by Southgate *et al.* (1957), but the two forms can be easily distinguished by their elytral colour pattern. Dimensional measurements were made on eighteen body structures of both forms and both sexes (Yasui, unpublished). Among these, most remarkable differences were found in body length, width between compound eyes, size of hind wing and length of pygidium at the adult emergence. It seems that each value of the difference implies the difference in locomotive activity, flight and walk, and that in sensory activity, visual and tactile.

On dissection, the abdominal cavity is filled with reproductive organs in the flightless form, especially in the female the ovary develops completely and each ovariole contains fully developed eggs. On the other hand, the abdominal cavity of the flight form is found to be filled with fat body. The ovary does not develop, and contains only a small number of undeveloped eggs. Bursa copulatrix is virtually smaller than in the flightless (Spirina, 1974). Several days after the emergence, the eggs begin to develop and a few number of fertile eggs is deposited throughout the whole life span. The testis of flight form is also immature or 'sterile' at the emergence, but it grows and matures gradually with the lapse of adult life (Gill *et al.*, 1971). The spermatozoa from this 'sterile' individual survive for less time and move more sluggish than those in the normal one (Bawa *et al.*, 1974).

Chemical composition of both forms were compared. Water content is low and crude fat content is high in the flight form (Utida & Takahashi, 1958). This lipid increase in the flight form is primarily due to triglycerides, which is the major part of lipids in both forms. No difference is found in the fatty acid composition between two forms (Nwanze *et al.*, 1976). Paper chromatography of the free amino acids in the male of both forms has not revealed any difference (Bawa *et al.*, 1974a).

Duration of the pre-imaginal development is about one and a half as long in the flight form as in the flightless form. Further, the flight form has several days of pre-maturation period after the emergence. Mean adult longevity in the flight form is about twice as long as in the flightless form. Its longest record was more than two months, in contrast with seven days in mean longevity of the flightless form. Thus, the total duration required for one life cycle is at least twice as long in the flight form.

Influences of temperature and atmospheric humidity on the life history phenomena, such as adult longevity, fecundity, fertility, etc., are different for the two forms. Generally speaking, the flight form is better able to survive at low temperatures and high humidities, and vice versa in the flightless form. The effects of crowding on the life history phenomena are more remarkable in the flight form.

In concluding, it can be said that each dimorphic character adapts well to the migratory life in fields and the population multiplication under the storage condition of beans, respectively.

Determination of the forms

For determining the two forms in the adult, two complemental factors, genetical and environmental, seem to be acting together. Although a genetical constitution is the same, the differentiation in adult forms is determined by the influences of environmental factors, such as larval crowding, temperature, water content of beans and photo-period during larval development. Among them, the most potent is the effect of larval crowding in the middle stage of its development, which may result into the processes of temperature rise in heap of beans (Sano, 1967) and change of water content of beans.

Physiological mechanism involved in the induction of flight form by these environmental factors is unknown, but a reduced level of juvenile hormone in the old larva or the pupa may be suggested (Utida, 1972; Nwanze *et al.*, 1976).

Genetical determination of the forms was indicated by the following findings. When a culture continued from generation to generation under the crowding condition, the percentage emergence of the flight form decreased gradually to a low level until after 45 generations it decreased to zero, though it kept over 30% at the start (Utida, 1974). The same decreasing trend was also found in the ordinarily maintained cultures in five successive years (Caswell, 1960). Backward process of selection of the flight form was made starting from the flightless population. In the resulted population after the lapse of generations, the percentage of the flight form attained high level, fluctuating 40 to 73. The reciprocal crosses between the 'start' and the 'selected' showed the intermediate percentage in the flight form emergence, as expected (Sano-Fujii, unpublished). These facts mentioned above imply the importance of genetical background in the expression of the two forms in adult. The flight form never appears in the end population of Utida's experiment, even if any environmental factor operates to produce it, and reversely in the Sano-Fujii's selected population it appears with ease without any effect of environment. Therefore, it can be concluded that the determination of the forms is the result of. interaction of genetical and environmental factors.

Here, it is worthy to mention that the dimorphic population seems to be not prevalent. Among 12 strains of the author's culture, those come from different geographical area, only two strains were dimorphic. The other 10 strains were flightless monomorphic one. Besides, it is said that the finding frequency of the flight form is also low on the occasion of plant quarantine inspection. These two kinds of population, dimorphic or monomorphic, may correspond to the populations selected experimentally by Sano-Fujii or Utida, to both extreme sides.

CONSIDERATIONS

The differences between two forms in *C. maculatus* are remarkable in their morphology, physiology, behaviour, etc., and each character is considered to be related to the resident in store and the migrant in field. In this connection, this dimorphism is analogous to the phase polymorphism in migratory locusts or the wing polymorphism in aphids. This analogy is more strengthened by the similarity in their induction mechanism.

As other pest species in *Callosobruchus*, *C. maculatus* has two phases in its life, one in the field on leguminous plant and the other in the warehouse on stored beans. The author reasoned that the two forms just correspond to these two phases of life. Number of generations will continue with the emergence of the flightless form in the store, and when the flight form emerges it will fly to a field where the female deposits the eggs on the pods of growing beans.

But, this inference contradicts partly the presence of flightless monomorphic strains, mentioned above. Some of these 'flightless' forms are very active in their behaviour, and are even good flyer. Further, the 'flightless' form could be seen in the same way in the field as the flight form (Taylor, 1970). If this is the case, the two phases of life, in field and storage, could be carried over fully by this 'flightless' form alone. Because the flight form should be replaced its situation in field by the 'flightless' form.

For the co-existance of both forms in an area, the flight form must repeat and complete its life cycle in the field alone, in separate from the cross life history of this 'flightless' form. It is considered that the necessary conditions for completing an annual cycle in the field may be the combination of the polyphagous habit and the prolongation of each life stage of the flight form, and the seasonal succession of seeds of host beans (wild or cultivated) in the field, in combined with the joint cultivation of early and late varieties of beans.

The place where these conditions are fulfilled may be found in the tropic now, and is assumed to be the origin or the distribution area in the age before the beans were cultivated and stored by human being. In the other area of distribution, we can only find the flightless monomorphic population.

Finally, the author dares to say that a similar dimorphism can be found in other species of *Callosobruchus*, when we make an effort to search along these lines.

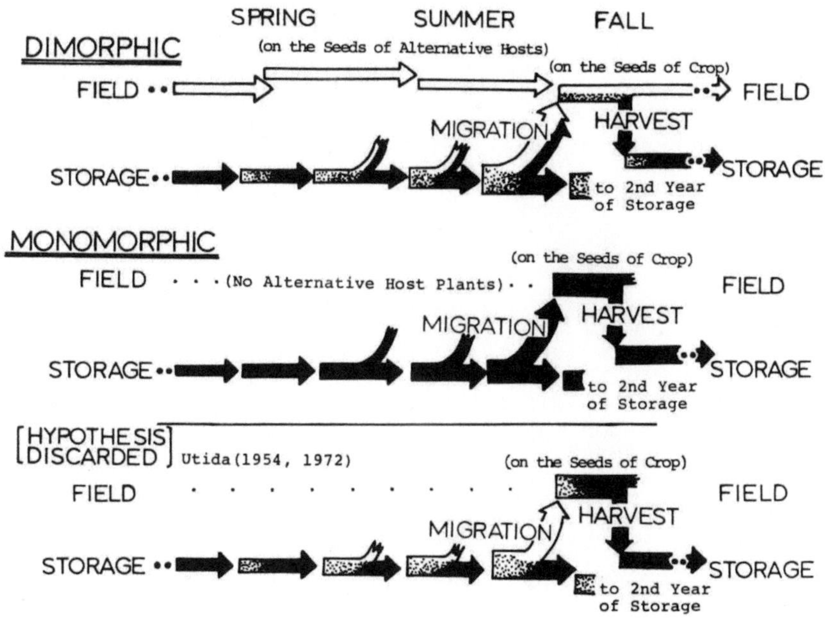

Fig. 1. Life histories of the two forms of *C. maculatus* alternating their generations in field and store (presumptive).
➡: flightless form, ⇨ : flight form.

REFERENCES*

Arora, G.L. & Pajni, H.R., 1959. Curr. Sci. 28: 19–20.
Bawa, S.R. *et al.*, 1974. Ann. Ent. Soc. Amer. 67: 519–520.
Bawa, S.R. *et al.*, 1974a. Ann. Ent. Soc. Amer. 67: 997–999.
Breitenbecher, J.K., 1921. Genetics 6: 65–90.
Caswell, G.H., 1960. Bull. Ent. Res. 50: 671–680.
Gill, J. *et al.*, 1971. Ann. Ent. Soc. Amer. 64: 1186–1188.
Kasiwagi, M. & Utida, S. 1972. Appl. Ent. Zool. 7: 95–96.
Nakamura, H., 1966. Jap. J. Ecol. 16: 236–241.
Nwanze, K. *et al.*, 1976. J. Insect Physiol. 22: 897–899.
Sano, I., 1967. J. Stored Prod. Res. 2: 187–195.
Southgate, B.J. *et al.*, 1957. Bull. Ent. Res. 48: 79–89.
Spirina, T.S., 1974. Ent. Rev. Wash. 53: 22–27.
Takenouchi, Y., 1971. Can. J. Genet. Cytol. 13: 708–713.
Takenouchi, Y., 1971a. Kontyu. 39: 332–337.
Taylor, T.A., 1970. Niger. Ent. Mag. 2: 66–69.
Utida, S., 1954. Jap. J. Appl. Zool. 18: 161–168.
Utida, S., 1972. J. Stored Prod. Res. 8: 111–126.
Utida, S., 1974. Proc. 1st Intern. Conf. Stored-Prod. Ent., pp. 686–691.

*Descriptions in the text not specified by author's name come from the present author; these source references were given in Utida (1972).

Author's address:
546-8 Issiki
Hayama
Kanagawa-ken, 240-01
Japan

REPRODUCTIVE POLYMORPHISM OF POPULATIONS OF *ACANTHOSCELIDES OBTECTUS* FROM DIFFERENT COLOMBIAN ECOSYSTEMS

J. HUIGNARD & J.C. BIEMONT
(*Tours, France*)

INTRODUCTION

The bean weevil *Acanthoscelides obtectus*, originates in the equatorial zones of South America (Hoffmann, 1945; Zacher, 1955). While the wild Phaseolinae that grow profusely in the area were its first host plants, its extension was subsequent to the domestication of *Phaseolus vulgaris*. For the cultivation of the bean, as it increases the available plant mass, also increases the possibilities of the insect encountering its host plant. The larvae stand a better chance of developing since bean selection increases the size of the pods and the volume of the seeds and a changed texture of pod pericarp and seed coats (Evans, 1980).

Acanthoscelides obtectus, then, has followed the extension of *Phaseolus vulgaris* crops in some areas, first in Latin America, then in Northern America, where the climatic conditions allow its development. The zones in which it lives have a temperate climate, since it is very susceptible to temperature variations. High diurnal temperatures – exceeding 35°C – sterilize the adults and stop larval development (Huignard & Biemont; Biemont, 1979).

No report of *Acanthoscelides obtectus* in Europe was made before the late 19th century, a good three hundred years after bean crops were introducted (Fabre, 1870). From an area in the south-east of France it colonized a large part of Europe, and was later able to spread to Africa following the introduction of *Phaseolus vulgaris* crops. As climatic conditions in Europe do not make for the insect's survival in nature in winter (Labeyrie, 1957) preservation of the insect populations is ensured by mans stored beans throughout the wintry season.

European populations, should derive from a few individuals; their genetic pool can be assumed to be smaller than that of populations living in the zone of origin. This made it interesting to study populations of *Acanthoscelides obtectus* from a zone of origin which, granting the biological characteristics of the insect – no diapause; thermosensibility – must be located in an equatorial area. With many wild or domesticated *Phaseolus* growing there, its geographic area (between 2 and 6°N) and the climatic conditions prevailing at some levels of the Andean cordillera, the central part of Colombia can be described as one of the areas from which *Acanthoscelides obtectus* may

Series Entomologica, Vol. 19: ed. by V. Labeyrie

originate. The topography of the area facilitates the geographic isolation of populations of *Acanthoscelides obtectus* since the three cordilleras, divided by many deep hot valleys (Fig. 1) prevent the development of *Acanthoscelides obtectus.*

We therefore visited in 1975 several provinces of the central part of Colombia from which we brought back various populations of *Acanthoscelides obtectus*. In most cases, these insects come from beans harvested one or two months before the collect; they probably originated from nature that is, from

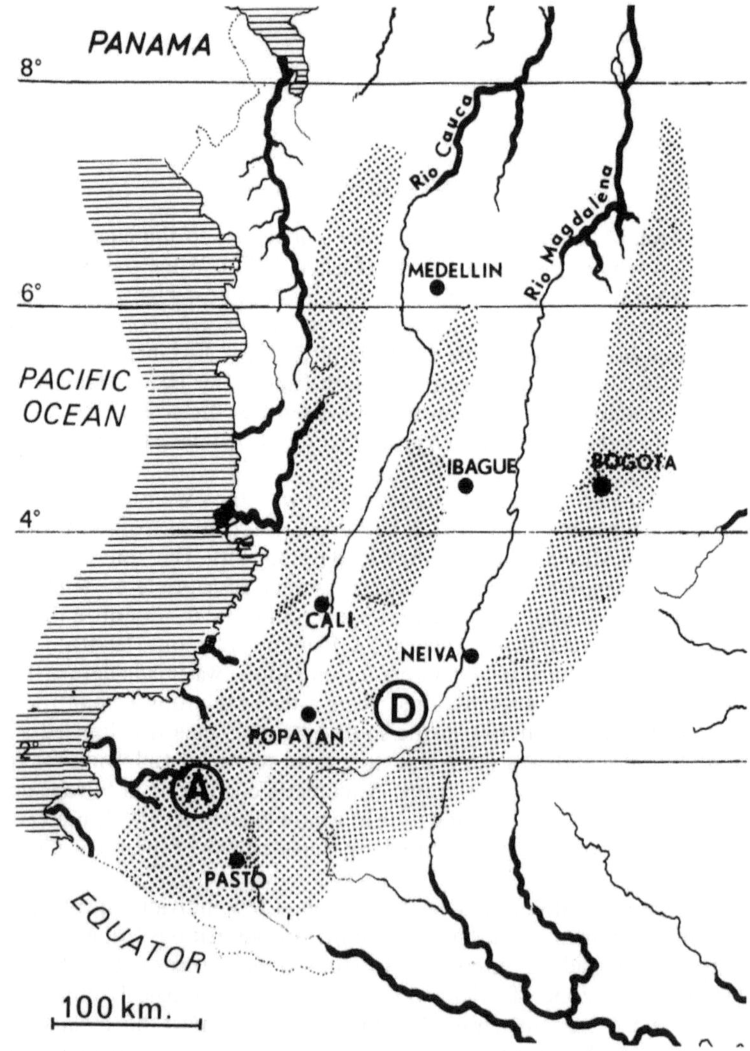

Fig. 1. Map of Colombia.
A = Origin area of the 2000 m strain (Narino)
D = Origin area of the 1000 m strain (Huila)

beans harvested in the field, where seeds had been infested. Bean crops in Colombia are grown either in the valleys or on the slopes of the cordilleras – up to 2000 m and over, depending on climate conditions.

In the lower areas – altitudes below 1000 m – where diurnal temperatures are very high (CIAT, publication 1976) *Acanthoscelides obtectus* causes any damages to the *Phaseolus vulgaris* crops. Extensive damage is caused by *Zabrotes subfasciatus*, another Bruchidae well adapted to these climatic conditions.

Most damage to *Phaseolus vulgaris* bean crops is caused by *Acanthoscelides obtectus* in areas extending from 1000 to 2000 m in altitude, whose mean diurnal temperatures range between 19 and 30 °C (CIAT publication 1976).

The weevil populations, collected in borderline areas, were the following:

– strains from low altitude areas (about 1000 m);
– strains from medium altitude areas (2000 m) – nocturnal and diurnal temperatures there are lower due to altitude and humidity.

The insects were reared in conditions similar to those of their original habitat, and their reproductive activity was compared in an attempt to understand the process of bean infestation.

BREEDING CONDITIONS – METHODS OF INVESTIGATION

The various strains have been reared in the laboratory for five years in climatic conditions best approximating those of their original habitat – the relevant data were provided by local weather reports (CIAT, publication 1976).

– *Strains from low altitude areas* (La Plata, Province of Huila)
  Photoperiod 12/12 (area between 2 and 4 ° NL)
  Thermoperiod 25 °C ± 2 (L)/15 °C ± 2 (D)
  Relative humidity 60 to 70% H.R. (L)/80% H.R. (D)
– *Strains from 2000 m areas* (Buesaco, Province of Nariño)
  Photoperiod 12/12 (area between 0–2 ° NL)
  Thermoperiod 21 °C ± 2 (L)/13 °C ± 2 (D)
  Relative humidity 60% to 70% H.R. (L)/80% H.R. (D)

Some adults were put soon after emergence in rearing boxes containing bean seeds; they then reproduced and laid eggs from which the next generation hatched out. Others were used for experimentation. Adults sexed soon after emergence were isolated in Petri dishes and fed a honeyed water.

During experimentation, the following criteria were examined:

– *Study of fecundity* (number of eggs released) and *ovarian production* (number of eggs laid + number of mature oocytes retained in lateral oviducts.)
– *Presence of vitellogenin in hemolymph*: vitellogenin studied by Biemont (1979) in *Acanthoscelides obtectus* is detected by Ouchterlony's immunodiffusion technique (1949) in whole female extracts tested with oocyte antiserum obtained by immunizing a rabbit with several injections of oocyte extracts emulsified with Freund's adjuvant (Fig. 2).
– *Histological study of ovarioles* based on sections fixed on alcoholic Bouin and stained with Masson Trichrome.

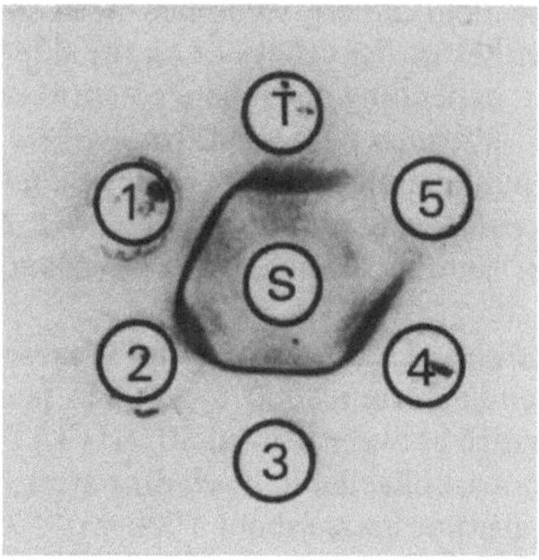

Fig. 2. Research of vitellogenin by immunodiffusion (Ouchterlony technique).
S = Anti egg serum. – T = egg extract. 1 to 4 = females vitellogenin + 5 = female vitellogenin –.

- *Determination of the level of juvenile hormones in hemolymph* and quantitative analysis of these hormones by radio-immunological test (Baehr 1977–1978).

RESULTS

Various factors – the presence of beans or ripe pods of *Phaseolus vulgaris*, or copulation –, being likely to influence the reproductive activity of virgin females of *Acanthoscelides obtectus* (Labeyrie, 1962; Huignard, 1975; Pouzat, 1978, Jarry, 1980) their action was tested succesively under the following conditions:

- reproductive activity of – isolated – virgin females;
- reproductive activity of virgin females in the presence of *Phaseolus vulgaris* beans or pods;
- reproductive activity of inseminated females in the presence of *Phaseolus vulgaris* seeds.

Female reproductive activities were analysed over several generations at different stages in the experiment.

REPRODUCTIVE ACTIVITY OF VIRGIN FEMALES

1. *Strains from the 2000 m areas (Province of Nariño)*

Females isolated just after emergence and fed a honeyed water diet live an average 100 to 150 days. During the imaginal life oviposition never occurred. Ovarian production is examined in different females at various stages of the experiment – from day 0 to day 150.

Results obtained along the first three generations (1975–1976)

Most virgin females produced no mature oocyte throughout the surveyed period. Their ovarioles (Fig. 3) only contained oocytes in previtellogenesis.

Observation of the basal oocyte shows that after a slight increase phase in the first few days of imaginal life growth is arrested, sometimes until the end of the experiment (Biemont, 1979).

Only a few (4 to 5%) females produced a small number of mature chorionated oocytes accumulating in lateral oviducts by the end of the experiment.

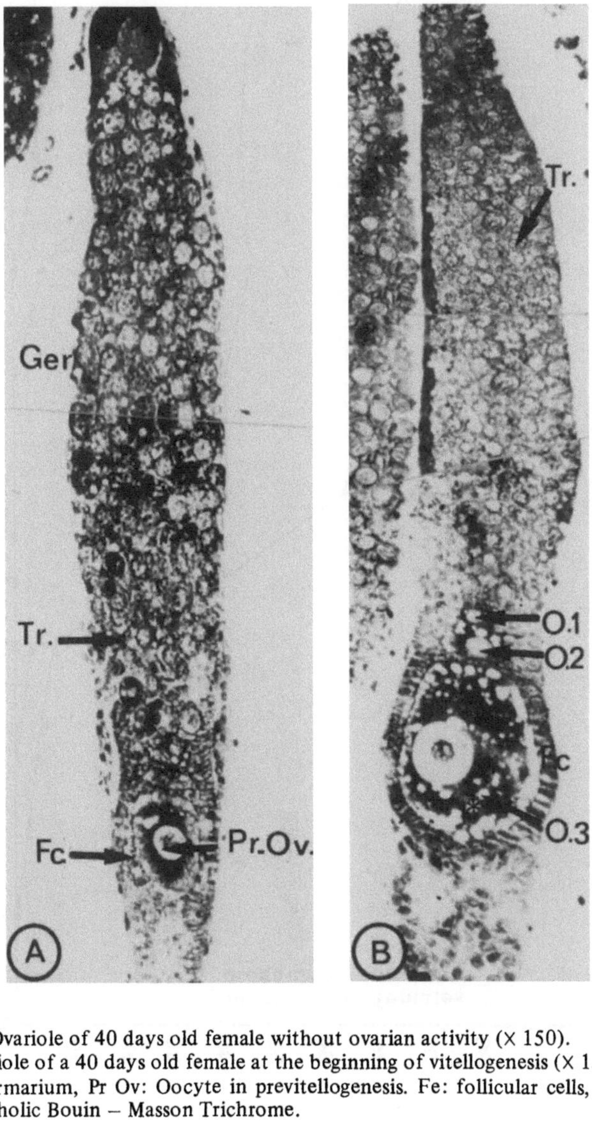

Fig. 3. A: Ovariole of 40 days old female without ovarian activity (X 150).
B: Ovariole of a 40 days old female at the beginning of vitellogenesis (X 150).
Ger: germarium, Pr Ov: Oocyte in previtellogenesis. Fe: follicular cells, Tr: trophocytes. Alcoholic Bouin – Masson Trichrome.

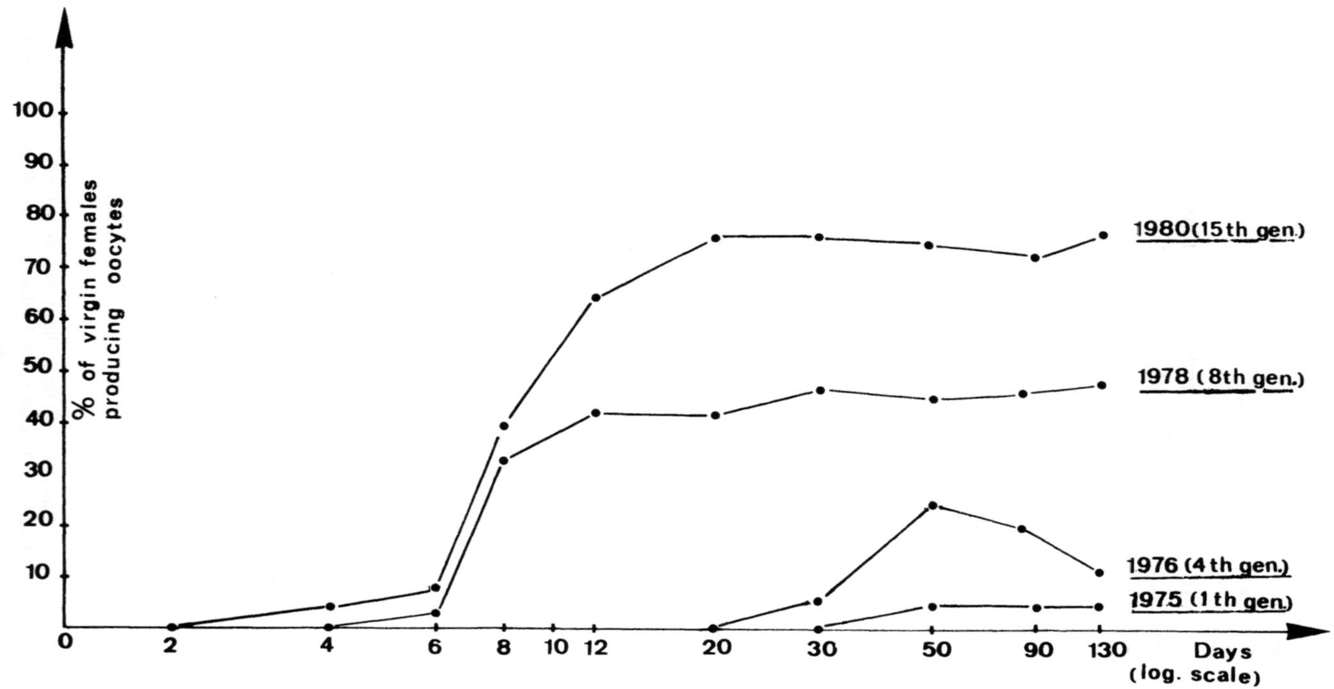

Fig. 4. Evolution of the percentage of females (2000 m strain) producing mature chorionated oocytes during 15 generations in the laboratory conditions.

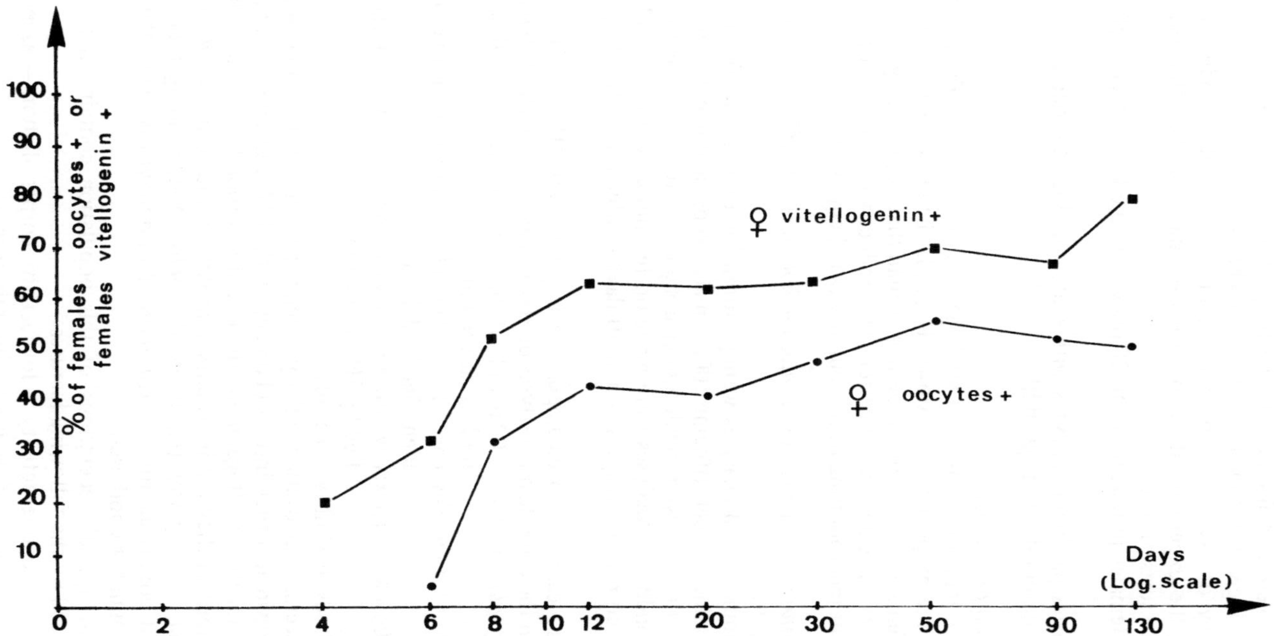

Fig. 5. Evolution of the percentage of females which synthesize vitellogenin and produce mature oocytes during the imaginal life (females of 2000 m strain).

Results obtained along the following generations (from 1977 to 1980)

1) Virgin females producing no mature chorionated oocyte in these conditions were still found, but the proportion of females with an ovarian activity increased along the generations (Fig. 4).
2) A survey of the evolution of the number of females producing oocytes during the experimental period leads to the conclusion that there are in fact two sorts of insects:
   either females producing 15 to 20 oocytes in the first few days of imaginal life;
   or females producing oocytes with a latent period of variable length in the different periods of imaginal life.

Vitellogenin production

20% of the females synthetized vitellogenin in the first few days of imaginal life, then the proportion rose slowly (Fig. 5). The proportion of vitellogenin-forming females was found to be higher than that of producing oocytes. One draws the conclusion that some females produce vitellogenin during imaginal life without incorporating it. It accumulates in hemolymph.

Radio-immunological analyses of juvenile hormones at different stages of the experiment.

Radio-immunological analyses were performed in each set on 100 μl of hemolymph obtained by centrifuging 1000 females whose abdomens were ventrally opened – the cuts were made under a stereo microscope so as to examine ovarian condition. Analyses were made on females aged 6 days – which had produced no mature oocytes – and 30 days – which had, or not, produced oocytes.*

In every case, it proved impossible to detect any JH_1 or JH_2 in hemolymph. On the other hand, analyses of JH_3 revealed significant differences:

– *6 day-old females*: 1.02 ng/ml of immunoreactive material identified by JH_3 anti-serum

– *30 day-old females*: (ov. –): 0.13 ng/ml of immunoreactive material identified by JH_3 anti-serum

– *30 day-old females*: (ov. +): 8 ng/ml of immunoreactive material identified by JH_3 anti-serum.

(methods sensibility: 1 ng/ml)

Thus, it seems that ovarian activity is related to the level of circulating JH_3, as was shown in other insects (Luscher *et al.*, 1976; Lanzrein *et al.*, 1978). One assumes that vitellogenin synthesis and incorporation are controlled by this hormone as they are in *Rhodnius prolixus* (Abu Akima *et al.*, 1977) or *Drosophila melanogaster* (Gavin *et al.*, 1976). Yet, granting that selective removal of corpus allatum in *Acanthoscelides obtectus* is impossible, one is left to formulate hypotheses.

To sum up, all generations of this population contain virgin females characterized by the following features:

– a low endocrine activity – low activity corpus allatum – as evidenced by the low level of circulating juvenile hormone;

*We thank Dr. J.C. Baehr for his help.

– the absence or the low level of vitellogenin in hemolymph, as was shown recently by quantitative techniques (Huignard & Leroi, 1980)
– previtellogenic ovarioles.

2. *Strains from the low atlitude areas*

Virgin females isolated as soon as they emerged from the bean seeds and fed a honeyed water diet had a 60 to 100 day imaginal life. They synthetized vitellogenin in the first few hours following imaginal exuviation.

– *prior to emergence*:	no female had any vitellogenin (non-melanized adults)
– *emergent adults*:	20♀/20 synthetize vitellogenin
– *adults 3 hours after emergence*:	16♀/20 synthetize vitellogenin
– *adults 16 hours after emergence*:	20♀/20 synthetize vitellogenin.

The first mature chorionated oocytes were present in lateral oviducts as early as the second day after emergence. Ovarian production gradually increased and reached a plateau about the 6th or 7th day. It was arrested from that time on when females were kept in these conditions, as exemplified in the European strains (Huignard, 1975).

3. *Host plant influence on the reproductive activity of virgin females*

A. *2000 m strain females*

1. *Virgin females reared in the presence of Phaseolus vulgaris seeds.* When the females were placed from emergence in the presence of bean seeds, results differed according to the generations:
– *earlier-generation females*: after an 8 to 10 days latent period, the females produced mature oocytes that were retained in the lateral oviducts or released – oviposition, small in every case, always occurred late in the experiment. A constant 10 to 15% of the females were not stimulated by *Phaseolus* seeds and produced no oocytes.
– *8–9th generations females*: The beans induced far quicker responses in this lot; these responses were (Table 1):
– a rapid synthesis of vitellogenin in the first few days of imaginal life;
– the formation of mature oocytes accumulating in the lateral oviducts and usually released – but with a high interfemale variability;
– as already evidenced in European weevils, exposure to the host plant stimulated oogenesis (Table 2). Ovarian production was always higher than that found in those virgin females isolated in absence of seeds and producing mature oocytes).

2. *Virgin females reared in the presence of Phaseolus vulgaris pods.* In these conditions the following observations were made 40 days later:
1) All females had produced mature oocytes.

Table 1. Reproductive activity of virgin females in presence of *Phaseolus vulgaris* seeds (strain of 2000 m altitude).

Age (days)	Number of females	% of females vitellogenin +	% of females with mature ooctyes
1	38	63	0
3	40	90	0
5	64	97	92
10	35	100	90
20	43	100	100

Table 2. Ovarian production of virgin females in presence or in absence of *Phaseolus vulgaris* seeds (strain of 2000 m altitude).

	Isolated virgin females producing mature oocytes (A)	Virgin females with bean seeds (B)	Signification of the difference A/B
20 days	n = 53 18.8 ± 2	n = 63 44.6 ± 2.5	t* = 15.3(+)
50 days	n = 48 15.6 ± 2.1	n = 41 30.3 ± 4.3	t = 5.9(+)
100 days	n = 47 18.2 ± 2.5	n = 21 58.3 ± 7.4	t = 10(+)

*Student t test

2) 25% of females lay eggs into the pods. The ventral pod suture was always chosen for that purpose.

3) The presence of the pod always stimulated oogenesis (as shown by the fact that ovarian production in these females was higher than that of the isolated virgin females that had formed oocytes). However, this stimulation looked weaker than that induced by the presence of *Phaseolus* seeds.

– ovarian production of 39 virgin females with pods: 30.7 ± 3.1

– ovarian production of 63 virgin females with seeds: 46.6 ± 2.5

(Mean significative difference $P < 0.001$)

These results were confirmed by a study of female distribution in relation with ovarian production (Fig. 6).

B. *Low altitude virgin females*

The presence of *Phaseolus vulgaris* seeds stimulated oogenesis and induced oviposition – but with a high interfemale variability. Results were similar on all counts to those obtained with European strains (Labeyrie, 1962; Huignard, 1970).

4. *Reproductive activity of mated females*

A. *2000 m strains (reared at 21 °C/13 °C)*

The females sexual receptivity was related to their ovarian condition whatever the duration of their contact with males (Huignard *et al.*, 1979). Females

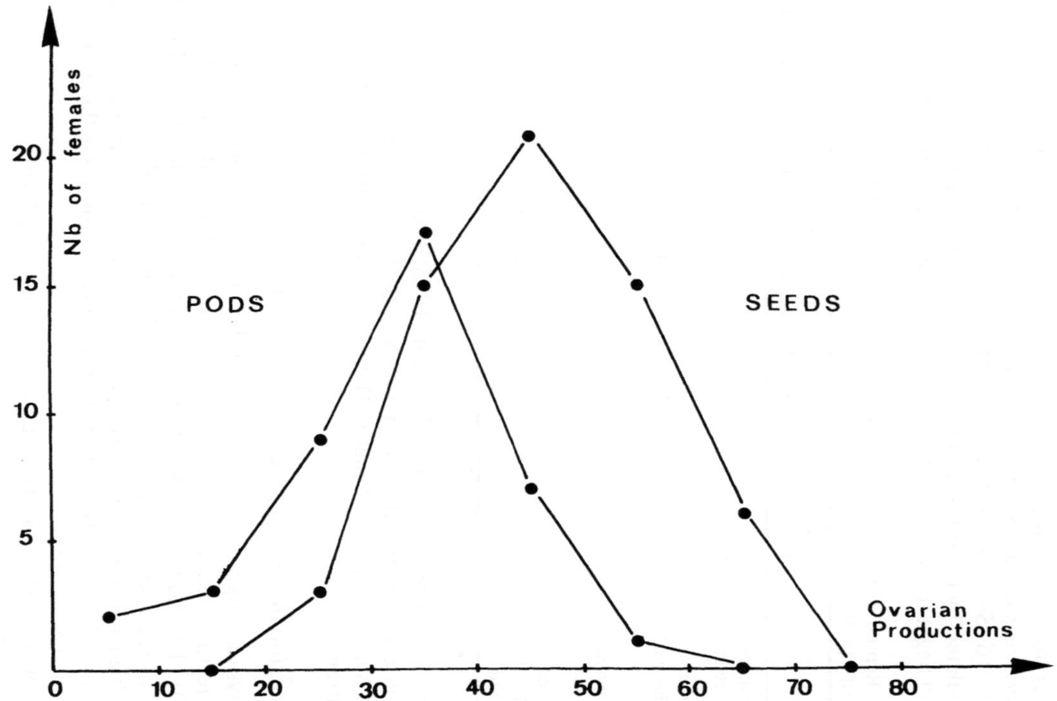

Fig. 6. Distribution of virgin females in relation with their ovarian production in presence of ripe pods or seeds of *Phaseolus vulgaris.*

accepted mating only when there were mature oocytes in the basal portion of the ovarioles or in the lateral oviducts.

Fecundity – ovarian production (Table 3). The females were put in the continuous presence of males as early as the 4th day of imaginal life; bean seeds or ripe pods were introduced in the rearing boxes as from the 10th day of imaginal life.

It was found no oviposition ever occurred before the introduction of *Phaseolus vulgaris* seeds or pods. The first ovipositions were observed as from day 16. Since in most females vitellogenin synthesis and incorporation are stimulated by the introduction of *Phaseolus vulgaris* seeds or pods, females mate and begin to oviposit only after the first mature oocytes are produced.

Table 3. Reproductive activity of inseminated females (2000 m strain) in presence of seeds or ripe pods.

Females	Number of Females	Fecundities	Ovarian productions	Signification of the difference
Virgin females with seeds (T_1)	38	Variable	39.5 ± 3.1	
Inseminated females with seeds	43	42.4 + 4.7	50.4 ± 4.9	t = 3.6 (+)(T.1)
Virgin females with pods (T_2)	39	variable	30.7 ± 3.1	
Inseminated females with ripe pods	36	36.1 ± 6.6	48.7 ± 6.7	t = 4.7 (+)(T.2)

As in the European strains (Huignard, 1975; Pouzat, 1978), host plant-mating interaction was the factor which strongly stimulated oogenesis and induced oviposition in all females. Mating alone failed to stimulate oogenesis in this strain.

B. *Lower altitude strains* (conditions 25 °/15 °C)

Four days-old virgin females – that had then already formed oocytes – were put in rearing boxes with males. Bean seeds were introduced 6 days later (day 10). Fecundity was surveyed every sixth day – until day 40, since females' lives are short in these conditions. The following data were found:

1) a high percentage of females (30 to 40%) oviposit in the absence of bean seeds. This capacity is virtually maintained throughout the generations (Huignard *et al.*, 1978).

2) Copulation stimulates oogenesis and induces egg laying:

Ovarian production of
48 virgin females + beans (24 days) 40.7 ± 2.7

Ovarian production of
52 mated females + beans (24 days) 76.1 ± 6.2

(values obtained at generation 4)

DISCUSSION

The experiments carried out in the present survey showed the existence of two allopatric populations of *Acanthoscelides obtectus* at various levels of the Andean cordillera in central Colombia. These two populations differed in their reproductive activities.

Females from medium (2000 m) altitude strains developed slowly and their imaginal life was long. Under experimental conditions, virgin females of the earlier generations produced no mature oocytes in the absence of host plant seeds or pods. The presence of seeds – or pods – stimulated vitellogenin synthesis and its later incorporation: acceptance of the males was subsequent to the production of a number of mature oocytes. This dependence of the females sexual receptivity on their ovarian condition has already been reported in *Musca domestica* (Adams *et al.*, 1969).

Lower altitude (about 1000 m) populations developed more rapidly, and all virgin females formed mature chorionated oocytes even in the absence of host plant seeds. As a consequence, oviposition can follow hard on copulation.

Differences between these two populations were maintained throughout the fifteen generations of the survey. However, the populations evolved as rearing proceeds so that medium-altitude virgin females 'emancipated' themselves from the influence of host plant seeds and reproductive activity was accelerated. This evolution was probably inherent to the rearing conditions which always placed mated females in the presence of bean seeds, that is, in conditions most likely to ensure the development of offspring.

The characteristics of these two allopatric populations also persist, at least in the earlier generations, when the strains are reared under conditions differing from those of their original habitat.

Two hypothesis are worth considering to explain the observed differences between these two allopatric populations:

1) The conditions of preimaginal development may affect imago physiology, and particularly the reproductive function as was shown, among others, in *Drosophila melanogaster* (Lints, 1971; Lints *et al.*, 1969). This hypothesis is not, however, supported by the data, since the insects from the various strains placed in climatic conditions unlike those of their original habitats maintain over the earlier generations a reproductive activity more or less identical to that observed under normal thermoperiodical rearing conditions (Huignard *et al.*, 1978).

2) Thermoperiodical conditions, which are different in those two systems, do not seem to affect directly the reproductive physiology of these two Bruchidae populations. They may yet affect them indirectly through their effect on the growth and reproduction of *Phaseolus vulgaris* whose pods induce the adult weevil's reproductive activity while the seeds are the larva's trophic basis. *Phaseolus* flowering and fructification are actually delayed by a lower temperature regime (Bierhuizen, 1970; Wallace, 1980). Thus, climatic conditions in the medium altitude area can be assumed to prevent the flowering of Phaseolinae, then the maturation of their seeds, during part of the year. Seeds suitable for the development of weevils may be available to them in

nature but over a few periods each year. Under these conditions, it is favourable for virgin females highly dependent on the host plant to undergo this type of ovarian diapause. This avoids a lengthy retention of mature oocytes in the lateral oviducts, a factor which may prove a disadvantage (Biemont, 1979).

The warmer equatorial climate prevailing in the lower altitude area allows *Phaseolus* to grow, flower and fructify the whole year round. Then the insects are continuously placed in conditions favouring their development. Selective pressures are eased off, as they will be in that case (Dobzansky, 1951; Southwood, 1976). Virgin females are able to produce oocytes in the absence of the host plant, and mated females can lay eggs even in the absence of seeds – which must lead to some not inconsiderable loss of eggs in nature, since the chances of the weak first instar larvae finding their trophic basis are nil. Actually, selective pressures being eased off in the laboratory too as a result of the rearing conditions, the evolution of the medium altitude strain along the generations is also accounted for.

Climatic conditions, by making host plant seeds less available, and natural barriers to communication between the ecosystems inhabited by the insects have led to the isolation in natural conditions of these allopatric, yet still interfecund, populations of *Acanthoscelides obtectus.*

Such research is relevant in agronomy for two reasons:

1) Efficient pre-harvest control of *Acanthoscelides obtectus* is dependent on collecting biological information on this insect in its natural environment. Now, bean fields may be inhabited at all times by a weevil population with the capacity to survive until the seeds are ripe and drying pods enable it to breed. Further study is needed to define the fate of these populations between breeding times; it will be concerned with finding the host plants likely to allow the development of the insects. Such fundamental research in biology falls into this perspective.

2) Weevil multiplication in stored beans can be curbed thanks to a survey of the climatic conditions favouring the development of *Acanthoscelides obtectus*, particularly if it revealed the insect's sensitivity to temperature. Combined with other techniques (Huignard, 1978), this perspective strategy is mostly valid in those areas where beans are infested by *Acanthoscelides obtectus* alone and not by *Zabrotes subfasciatus*, which is much less affected by high temperatures.

REFERENCES

Abu Hakima, R. & Davey, K.G., 1977. The action of juvenile hormone on follicle of *Rhodnius prolixus* in vitro: the effect of Colchicine and Cytochalasin B. Gen. Comp. End. 32: 306–370.

Adams, T.S. & Hintz, A.M., 1969. Relationships of age avarian development and the Corpus allatum to mating in house fly *Musca domestica.* J. Insect. Physiol. 14: 983–993.

Baehr, J.C., 1977. Aspects méthodologiques et critique du dosage radioimmunologique des hormones juveniles. Bull. Soc. Zool. Fr. 102: 3–300.

Baehr, J.C., 1978. Dosages radioimmunologiques des ecdysteroides et des hormones juveniles au cours des deux derniers stades larvaires de *Rhodnius prolixus.* C.R. Acad. Sci. Paris 287: 523–526.

Biemont, J.C., 1979. Vitellogenesis in *Acanthoscelides obtectus.* I: Oocyte development and vitellogenin in a european strain. Int. J. Invert. Reprod. 1: 221–232.

Biemont, J.C., 1979. Influence de la plante hôte et de la copulation sur la levée de l'inhibition du developpement ovarien lié à la rétention des ovocytes chez *Acanthoscelides obtectus.* Ann. Soc. Ent. Fr. 15 (1): 93–99.

Bierhuizen, J.F., 1970. The effect of temperature on plant growth and yield Proc. Uppsala Symp. UNESCO, 89–98.

CIAT publications (1976) Palmir Cuadro 2–3.

Dobzansky, T.T., 1951. Genetics and the origin of species. Columbia University Press.

Evans, A.M., 1980. Structure, variation, evolution and classification. Advances in Legume Science, pp. 337–347. R.J. Summerfield and A.M. Bunting.

Fabre, J.H., 1870. La bruche des haricots. Souvenirs entomologiques. 8ème Serie, 50–69.

Gavin, J.A. & Williamson, J.H., 1976. Juvenile hormone induced vitellogenesis in apterous a no vitellogenic mutant in *Drosophila melanogaster.* J. Insect Physiol. 22: 1737–1742.

Hoffmann, A., 1945. Coleopteres Bruchidae et Anthribides, Faune de France 44: 90–92, Le Chevalier.

Huignard, J., 1970. Influences comparées de la plante hôte et de la copulation sur l'ovogenèse et la ponte d'*Acanthoscelides obtectus* CR. Acad. Sci. Paris, 271: 2171–2174.

Huignard, J., 1974. Influence de la copulation sur la fonction reproductrice femelle chez *Acanthoscelides obtectus.* I: Copulation et spermatophore Ann. Sci. Nat. 12–16, 361–434.

Huignard, J., 1975. Influence de la copulation sur la fonction reproductrice femelle chez *Acanthoscelides obtectus* II: Analyse expérimentale. Ann. Sci. Nat. 12–17–I–66.

Huignard, J. & Biemont, J.C., 1978. Comparison of four populations of *Acanthoscelides obtectus* from different ecosystems. Assay of interpretation. Oecologia 35: 307–318.

Huignard, J., 1978. Regulation of the bean weevil reproduction *Acanthoscelides obtectus* and research on techniques of protection of stored beans. Proc. of the II Int. Work. Conf. St. Entomol. Ibadan. 360–371.

Huignard, J., & Biemont, J.C., 1979. Vitellogenesis in *Acanthoscelides obtectus* II) The conditions of vitellogenesis in a strain from Colombia. Int. J. of Invert. Reprod. I: 221–232.

Huignard, J. & Leroi, B., 1980. Analyse quantitative du taux de vitellogénine dans l'hémolymphe dans une lignée de bruches (*Acanthoscelides obtectus*) d'origine colombienne (submitted for publication to Experientia)

Jarry, M., 1980. Comparaison de fécondités et des pontes d'*Acanthoscelides obtectus* en présence de graines ou de gousses de *Phaseolus vulgaris.* Ann. Soc. Ent. Fr. 16 (3) 345–349

Labeyrie, V., 1957. Sur les conditions de pullulation en France de la bruche du haricot (*Acanthoscelides obtectus*) en culture. C.R. Acad. Agric. 43: 590–593.

Labeyrie, V., 1962. Mise en évidence d'influences multiples de la plante-hôte sur la stimulation de la ponte chez *Acanthoscelides obtectus.* C.R. Soc. Biol. 256: 8–9, 1473–1477.

Labeyrie, V., 1962. *Les Acanthoscelides* in Entomologie appliquée à l'Agriculture. 1: 4343–442 Masson.

Lanzrein, B., Gentinetta, V., Fehr, R. & Luscher, M., 1978. Correlation between haemolymph juvenile hormone titre, corpus allatum volume and corpus allatum in vivo and in vitro activity during oocyte maturation in a cockroach *Nauphoeta cinerea.* Gen. and Comp. Endocrinol. 36: 339–345.

Luscher, M. & Lanzrein, B., 1976. Differential effects of three juvenile hormones (JH I, II, III) in the cockroach *Nauphoeta cinerea.* Coll. C.N.R.S. 251 Villeneuve d'Asq. 436–440.

Lints, F.A., 1971. Life span in Drosophila. Gerontologia, 17: 33–55.

Lints, F.A. & Lints, C.V., 1969. Influence of preimaginal environment on fecondity and ageing in *Drosophila melanogaster* hybrids. I) Preimaginal population density. Exptl Geront. 4: 231–244.

Maingaud, D., 1899. Le Bruchus irresectus (Fahr) parasite des haricots cultivés. Bull. Soc. Et. Sc. Nat. Nimes 27: 103–107.
Ouchterlony, O., 1949. Antigen – Antibody reactions in gels. Acta Pathol. Microbiol. 26: 507–515.
Pouzat, J., 1978. La régulation de l'activité reproductrive des femelles de la bruche du haricot: influence de quelques facteurs externes (plante hôte-activité locomotrice). Thesis University of Tours, 171 pp.
Southwood, T.R.E., 1976. The stability of the trophic milieu its influence on the evolution behaviour and responsiveness to trophic signals. Coll. C.N.R.S. Tours, 455–476; C.N.R.S.
Wallace, D.H., 1980. Adaptation of *Phaseolus* to different environments. In: Advances in legume science, 349–358. R.J. Summerfield, and A.H. Buting.
Zacher, F., 1955. Verbreitung und Nahrplanzen der Speisebohnen *Acanthoscelides obtectus*. Mittlg. Dtsch. Ges. 14: 3–4.

Authors' address:
I.B.E.A.S. (ERA CNRS 328)
Université F. Rabelais
Parc de Grandmont
37200 Tours
France

INFLUENCE OF THE TEGUMENT OF *PHASEOLUS VULGARIS* SEEDS AND OF LARVAL DENSITY ON THE DEVELOPMENT OF *ACANTHOSCELIDES OBTECTUS* SAY

D. STAMOPOULOS & P.DESROCHES
(*Thessaloniki, Greece/Tours, France*)

INTRODUCTION

The larvae of *Acanthoscelides obtectus* Say develop only in a limited number of Phaseolinae seeds, such as *Phaseolus vulgaris* and *Phaseolus coccineus*.

This specificity would be essentially due to the secundary compounds (protease inhibitors, phytohemagglutinines, heteropolysaccharides . . .) present in the various parts of legume seeds. Every Bruchidae species would be able to detoxify only a small number of these compounds (Applebaum, 1964; Applebaum & Guez, 1972; Janzen, 1977).

Acanthoscelides obtectus larvae which hatch inside the pods, must bore the tegument of the seed before reaching the cotyledons. The first holes bored are then often used by a succession of several larvae (Labeyrie, 1960). The tegument can operate not only as a mechanical barrier, but also as a chemical one (Janzen, 1978) whose action can get added to that of cotyledons to inhibit larval development.

Acanthoscelides obtectus undergoes all of its larval development and pupation in one seed, however overpopulated that seed may be (Herford, 1936; Larson & Fisher, 1938) though the larvae of other species, certain *Acanthoscelides* in particular, change seed in the course of their development (Center & Johnson, 1974). In *Acanthoscelides obtectus*, overpopulation then causes a miniaturization of a certain number of individuals (Herford, 1936; Larson & Fisher, 1938; Lerin, 1975). In that species, the tegument therefore represents a barrier preventing larvae from escaping, since larvae have been observed to escape when they develop on seeds whose tegument has been removed, or on pills made from cotyledons.

In spite of the importance of this role played by the tegument, no study on *A. obtectus* has ever examined the particular role of that structure upon the larval development and apart from Lerin's work (1975), the consequences of larval density on the possibilities of development have not been studied. That is the reason why we will successively examine: The larval mortality caused by the tegument in the course of penetration or when it is artificially mixed with the cotyledon in the meal, or by overpopulation; Then we will study the effects of overpopulation upon the rate of development and weight of adults.

Series Entomologica, Vol. 19: ed. by V. Labeyrie

All the experiments were conducted with *Phaseolus vulgaris* L. seeds (Soisson variety) collected in September of the year of experiment and stored from December on in a chamber about 5 °C constant. The insects came from a strain collected in France and maintained in the laboratory over twenty generations on *P. vulgaris* seeds (Temperature: 25 °D/15 °L, Photo-periode: 12/12, R.H. 75%).

Experimental imperatives did not allow all the experiments to be conducted simultaneously and in the same chambers.

INFLUENCE OF TEGUMENT UPON PENETRATION

In seeds removed from the pods, larvae nearly always bore the side of the seed that is in contact with the bottom of the chamber.

So, in order to study penetration, seeds were placed separately in chambers about 1.7 cm^3, laying on their larger side. In half the experiments the seed was artificially bored with a needle about 0.2 mm in diameter, two holes to the lower side and two holes to the upper side. One or two fertile eggs were put into the chamber, which was closed with 'parafilm'

A: 1 egg + 1 unbored seed (n = 105)
B: 1 egg + 1 bored seed (n = 105)
C: 2 eggs + 1 unbored seed (n = 105)
D: 2 eggs + 1 bored seed (n = 99)

At the end of experiment were noted

- the larval mortality before penetration with or without boring attempted.
- the larval mortality after penetration into the seeds and the number of emerged adults.

Though the 4 experiments (Table 1) were conducted simultaneously in the same chamber, the frequency of hatchings was significantly different ($X^2 = 9.33$).

In the absence of a previous boring (A and C), mortality before penetration was very high, either with one or two larvae (0.77 and 0.71) but in both cases about half of them tried to bore the tegument (0.51 and 0.53).

On the other hand when the seed had been bored beforehand (B and D) most larvae penetrated into the seed. (0.62 and 0.86). When there were two larvae (C and D) one third of them used the same hole, either when it had been artificially bored or when it was bored by either of them.

Where pills of agglomerated extracts were used for larval development (Stamopoulos & Huignard, 1980) the lack of structured tegument did not prevent larval penetration but on the contrary made it easier since mortality outside was always very low, from 0.08 to 0.05 (Table 2).

INFLUENCE OF TEGUMENT ARTIFICIALLY MIXED WITH THE DIET

When the artificial diet in the form of pills contained the finely grounded tegument mixed with the cotyledon meal, larval mortality was almost total (Stamopoulos & Huignard, 1980)(table 2). Larvae seemed to die the earlier as the particles were finer.

Table 1. Effect of the *Phaseolus vulgaris* seed-coat on penetration of *Acanthoscelides obtectus* larvae.

Treatments	Total number of eggs laid	Number of hatched eggs	Larval mortality before penetration: without penetration attempt	Larval mortality before penetration: with penetration attempt	Larval penetration through seed-coat (with respect to hatched eggs)	Larval mortality after penetration (with respect to number of larvae that have penetrated the seed)	Emerged adults (with respect to hatched eggs)	Frequency of multiple utilisation of the same hole by the larvae
A. 1 egg + 1 unbored seed	105	78 (74%)	20 (26%)	40 (51%)	18 (23%)	2 (11%)	16 (20%)	–
B. 1 egg + 1 artificially bored seed	105	74 (70%)	16 (22%)	12 (16%)	46[a] (62%)	8 (17%)	38 (51%)	–
C. 2 eggs + 1 unbored seed	210	209 (99%)	38 (18%)	111 (53%)	60 (29%)	19 (31%)	41 (20%)	20 (33%)
D. 2 eggs + 1 artificially bored seed	198	185 (93%)	8 (4%)	18 (10%)	159[b] (86%)	38 (24%)	121 (65%)	53 (33%)

[a] 5 of them did not use the previously bored holes but bored their own.
[b] 7 of them did not use the previously bored holes but bored their own.

Table 2. Acanthoscelides obtectus development, into pills formed from different parts of *Phaseolus vulgaris* seeds (var. Soisson).

Contents of pills	Number of pills used	Number of eggs laid	Number of hatched eggs	Larval penetration (with respect to hatched eggs)	Emerged adults ♂	Emerged adults ♀	Larval mortality in the pills with respect to the larval penetration (%)	Mean larval development (days)	Mean weight of adults ♂	Mean weight of adults ♀
I. Cotyledon powder only	35	350	295	281 (95.2%)[a]	96	134	18.1[a]	67 ± 1[a]	4.7 ± 0.3[a]	6.3 ± 0.3[b]
II. Cotyledon powder + 10.2% seed-coat powder	35	350	302	281 (93.0)[a]	2	2	98.5[b]	92[b]	4.4	5.4
III. Whole seed powder	35	350	297	273 (91.9)[a]	2	3	98.1	94[b]	4.3	5.2

Figures not followed by the same letters are significantly different at $P = 0.05$.

To determine precisely the origin of this toxic effect, a tegument homogenate variously treated was introduced into the pills which still contained a similar quantity of cotyledon powder.

As Martin-Tangy *et al.*, (1977) have shown the presence of polyphenols in the tegument of *Vicia faba*, the homogenate was treated by shaking in various solvents (Martin-Tangy, personal communication):

A: For 4 h by H_2O at 80 °C
B: For 2 h by ethyl acetate (at 45 °C)
C: For 4 h by absolute methanol (at 55 °C)
D: For 3 h by methanol + H_2O (1 : 1)(at 60 °C)

To eliminate the action of heat labile allelochemical substances, (lectines, protease inhibitors, etc.) the tegument homogenate was sterilized for 20 min at 135°C (Janzen *et al.*, 1976); in all cases, toxicity remained very high (Table 3).

Table 3. Treatment of the seed-coat of *Ph. vulgaris* (Var. Soisson) with various solvents and autoclavage in order to eliminate toxic substances.

Treatment	Nb of pills used	Nb of eggs laid	Nb of hatched eggs	Nb of larval penetration	Nb of adults emerged
With methanol	30	120	102	87	0
With methanol + H_2O (1 : 1)	30	120	95	92	0
With ethyl-acetate	30	120	97	91	0
With H_2O	30	120	81	75	0
With Autoclavage	30	120	93	91	0
Tegument untreated (control)	30	120	94	84	0

INFLUENCE OF OVERPOPULATION

To examine the influence of overpopulation, groups (E.F.G.H.) each corresponding to about 20 replicates and containing 6, 12, 24 and 40 fertile eggs from females the same age, were placed on isolated seeds. The first three series were made simultaneously, but owing to the importance of the required number of eggs, the fourth series had to be performed two months later. The seeds in the first three series weighed 420 to 440 mg each, those in the last one about 370 mg. To make penetration through the tegument easier, 5 to 6 holes were bored into the seeds; in spite of that, the penetration rates were variable. To get elements for comparison, only the groups in which over 80% larvae had penetrated, were analysed.

As larvae do not necessarily occupy the seeds at the same time in a fifth series (J) two batches of 12 fertile eggs were introduced at an interval of 13 days on the same seeds weighing about 430 mg. Lastly to determine precisely the influence of starvation on the weight of adults, groups of larvae were taken out of the seeds at different instars and placed in Petri dishes without any food supply (Stamopoulos, 1978).

Whatever the larval density was, there was no significant alteration of the sex ratio in the adults obtained. Larval mortality after penetration (Table 4) was the same in groups E, F and G, and very low; on the other hand, it was higher in group H, but then larval density was increased in it by the lower weight of the seeds; it was therefore difficult to evaluate separately the influence of the increased larval number and of the reduced weight of seeds. The same numbers of adults were yielded per seed in group G and H. In group J the different times of introduction of the eggs allowed to make a difference between the adults that emerged from the two batches, and to show that the increased mortality rate, as compared to that of group G, was only due to the second batch in which it was over 80%.

The weight of adults was not affected by larval density up to 12 larvae per seed (E and F); the weights within each sex group were homogeneous, with females being heavier than males (Fig. 1), in accordance with usual observations. Groups G and H presented distribution with a translation of the curve (in comparison with E and F) and for G a lowering towards the lower values.

The differences between the weight of males and of females decreased in Group G and disappeared in group H. As the sex ratio remained identical, that means miniaturization affected females more than males. The translation of curve G with regards to E and F could mean that all insects were affected by weight reduction. The observations concerning mortality in the course of that experiment were entirely borne out by the weight of adults of group J. It could be noticed that in experiment G, larval density already induced a very important miniaturization (with females and males below 2 mg), whereas it had not caused increased mortality in the course of larval development.

Table 5 shows that since the third instar's larvae could reach the adult stage in spite of starvation. The adults thus obtained were miniaturized and their weight was comparable to those resulting from the overpopulation of seeds in experiments H and J. (Table 4).

The mean duration of development were similar for both sexes in all the groups of the experiment, with males emerging on the average one day before females (Table 4). The analysis of developmental duration in experiment G, with respect to the weight of adults (Fig. 2) showed a slight inverse correlation between these two parameters. That phenomenon can be observed in all experiments which can account for the mean developmental duration being practically identical.

CONCLUSION

The tegument of *Phaseolus vulgaris* seeds which have been ripe for two months or more, does not only act as a mechanical barrier, but it also contains toxic substances that cannot be degraded by all *A. obtectus* larvae.

Before making inferences from it on the poor adaptation of *A. obtectus* larvae to these protection of *P. vulgaris* seeds, it would be necessary to see if similar results can be obtained on other varieties, especially on those that most resemble the original plants. Likewise it would be necessary to examine whether that barrier appears as soon as pods ripen.

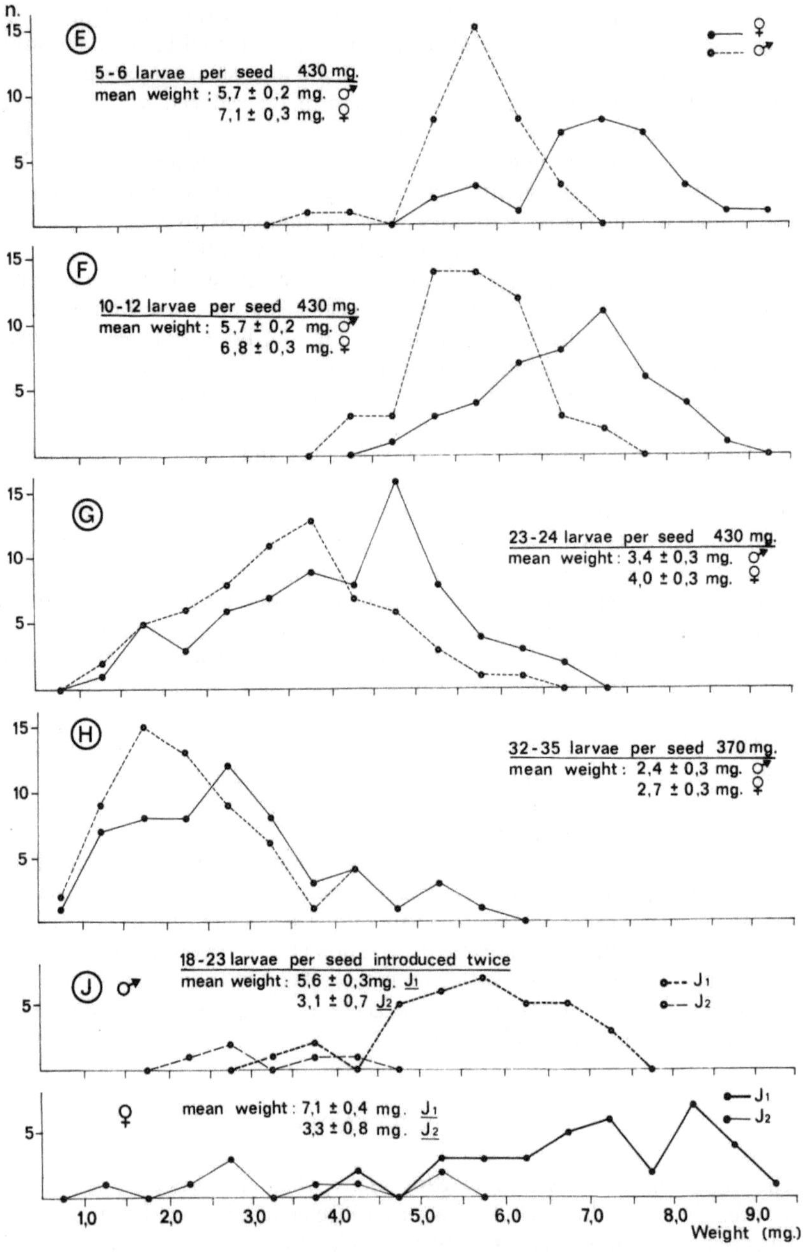

Fig. 1. Weight of newly emerged adults at different larval densities of *Acanthoscelides obtectus.*

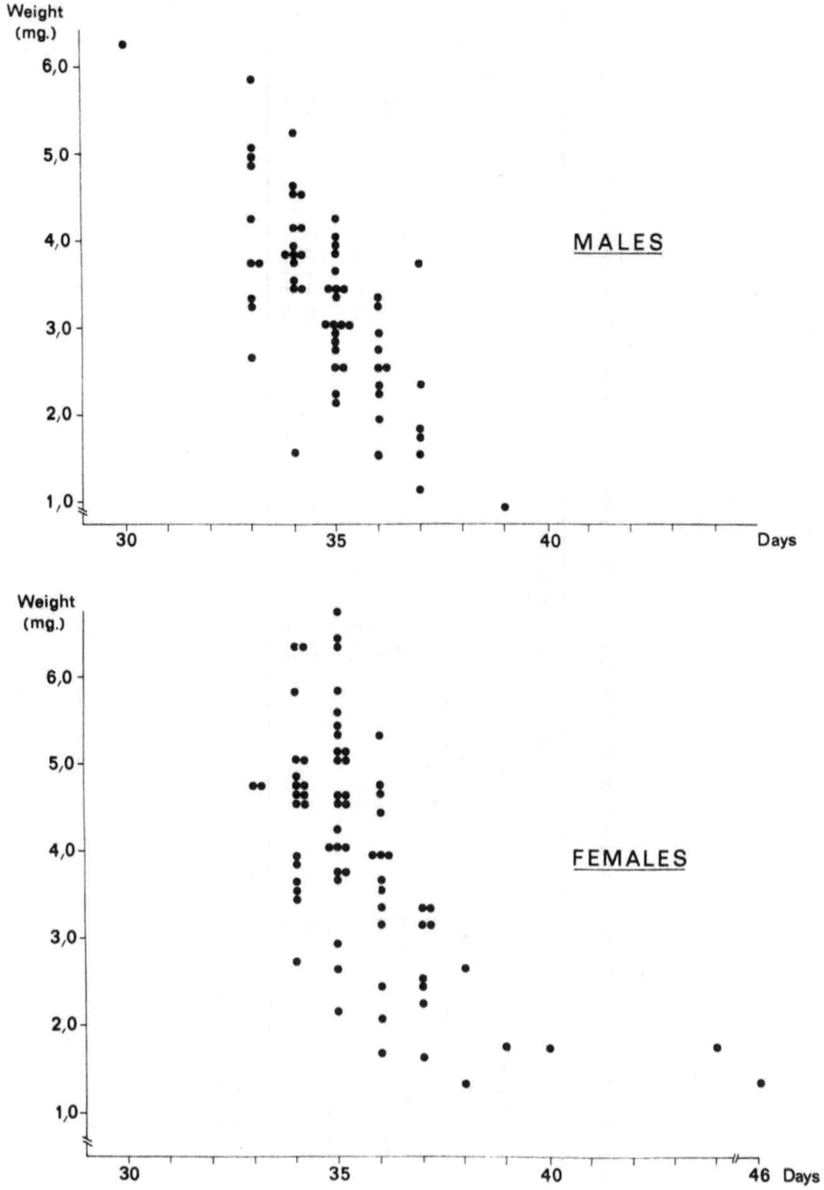

Fig. 2. Weight distribution of newly emerged adults with respect to length of larval development of *Acanthoscelides obtectus*. (Exp. G: 23–24 larvae per seed.)

Table 4. Duration of embryonic and post-embryonic development at different larval densities of *Acanthoscelides obtectus* (temperature: 26 ± 1 °C; R.H.: 70%).

Number of replicates	Larval penetration		Mean duration of development				Mortality inside cotyledons (%)
			Males		Females		
	per seed[a]	total	n	t (days)	n	t (days)	
(E) 14	5– 6	79	36	35.4 ± 0.5	33	36.6 ± 0.5	13 ± 7
(F) 10	10–12	109	51	35.5 ± 0.4	45	36.5 ± 0.5	12 ± 6
(G) 7	23–24	164	63	34.8 ± 0.4	73	36.6 ± 0.5	17 ± 6
(H) 6	32–35	199	64	36.4 ± 0.4	56	37.2 ± 0.5	40 ± 7
(J) 8	18–23	163					44 ± 7
J1[b]	11–12	91	42	35.7 ± 0.6	36	36.4 ± 0.7	14 ± 7
J2[b]	7–11	72	5	36.6 ± 2.0	9	36.3 ± 0.9	81 ± 9

[a] Seeds weighing about 430 mg (series E.F.G.H.) and 370 mg (J.)
[b] Batches J1 and J2 introduced at an interval of 13 days on the same seeds.

Table 5. Influence of larval starvation on the adults of *Acanthoscelides obtectus* (after Stamopoulos, 1978).

Instar at which larval starvation begins*	Number of replicates	Emerged adults		Weight of adults (mg)	
		Males	Females	Males	Females
L_2	36	0	0	–	–
L_3	48	12 (0.25)	2 (0.04)	2.7 ± 0.4	3.0
L_4	69	26 (0.38)	25 (0.36)	4.6 ± 0.5	5.3 ± 0.6
Control	–	30	30	4.8 ± 0.2	6.2 ± 0.1

*Starvation by taking larvae out of the seeds.

In effect larval mortality increases considerably on old seeds; therefore it cannot be excluded that in the early days following the ripening of pods, the tegument can be more easily bored by *A. obtectus* larvae. The rapid strengthening of the barrier after ripening would then become beneficial to that insect, by removing the hazards of escape and dessication of larvae and pupae outside the seeds.

If on the contrary the results obtained could be observed on all varieties and whatever the age of seeds since ripening, it could be inferred that *A. obtectus* is poorly adapted to the consumption of *Phaseolus vulgaris* seeds.

Only the importance of *Phaseolus vulgaris* cultivation could account for the high level of *Acanthoscelides obtectus* attacks on that host which should consequently be considered as a secundary one. Then the behavioural gregariousness of penetration should not be examined independently of the clustering ovipositional behaviour (up to 30 and 40 eggs at a time) and of the remarkable adaptation of larvae to high densities. In effect, when examining the ponderal yields (overall weight of adults related to the weight of seeds) from experiments E, F, G and H, the highest yield ($r = 0.17$) was shown to be obtained with 23–24 larvae per 430 mg seed (exp. G) and the one obtained with 32 to 35 larvae per 370 mg seed (exp. H) was still comparable ($r = 0.14$) to the one obtained with 10 to 12 larvae per 430 mg seed (exp. F). Thus the survival capacity and the efficient exploitation of the seed with a high larval density, the behavioural gregariousness of penetration and the clustering ovipositional behaviour, would be three consistent elements of the same adaptation to a plant species with large but rather scarce seeds which dry rather rapidly after ripening. In effect, the analysis showed (exp. J) that on the same seed, a second batch of hatchings did not allow larvae to settle in good conditions, as they were not provided with the same developmental possibilities they would get with a simultaneous hatching and a similar larval density.

However, interesting on the biological viewpoint these remarks on the effect of tegument and larval density may be, these observations also showed that this tegument contains an unknown factor of mortality, different from those that have already been discovered in legume seeds. On the other hand it does not seem to be certain that the utilization of trap-plants which would cause successive egg-laying to concentrate, by an earlier ripening, would allow the infestation of the other plants to be prevented.

REFERENCES

Applebaum, S.W., 1964. Physiological aspects of host specificity in the Bruchidae. I: General considerations of development compatibility. J. Insect. Physiol. 10: 783–788.

Applebaum, S.W. & Guez, M., 1972. Comparative resistance of *Phaseolus vulgaris* beans to *Callosobruchus chinensis* and *Acanthoscelides obtectus* (col. Bruchidae): The Differential digestion of soluble Heteropolysaccharide. Ent. Exp. & Appl. 25: 64–74.

Center, T. & Johnson, C., 1974. Coevolution of some beetles (Coleoptera: Bruchidae) and their hosts. Ecology 55: 1096–1103.

Herford, G., 1936. Observations on the biology of *Bruchus obtectus* Say with special reference to the nutritional factors. Z. Ang. Entom. 22: 26–50.

Janzen, D., 1977. The interaction of seed predators and seed chemistry. Coll. Int. C.N.R.S. 265: 415–418.

Janzen, D., 1978. The ecology and evolutionary biology of seed chemistry as relates to seed predation (in biochemical aspects of plant and animal coevolution). Harborne Academic Press, London, pp. 163–206.

Janzen, D., Juster, H. & Liener, J., 1976. Insecticidal action of the Phytohemagglutinin in black beans. Science, 192: 795–796.

Labeyrie, V., 1960. Nature de la répartition des larves d'*Acanthoscelides obtectus* say. C.R. Acad. Sci. 251: 149–151.

Larson, A.D. & Fisher, C.K., 1938. The bean weevil and the Southern cowpea weevil in California. Tech. Bull. No. 593, U.S. Department of Agriculture, 70 pp.

Lerin, J., 1975. Aspect et conséquences de la compétition larvaire lors de surpeuplements chez *Acanthoscelides obtectus* Say (Col. Bruchidae). Thèse de 3e cycle, Tours, 61 pp.

Martin-Tangy, J., Guillaume, J. & Kossa, A., 1977. Condensed tannins in horse bean seeds: chemical structure and apparent effects on poultry. J. Sc. Fd. Agric. 28: 757–765.

Stamopoulos, D., 1978. Influence de quelques facteurs agissant sur le développement embryònnaire et la reproduction d'*Acanthoscelides obtectus.* D.E.A. Université de Tours, 37 pp.

Stamopoulos, D. & Huignard, J., 1980. L'influence de diverses parties de la graine de haricot (*Phaseolus vulgaris*) sur le développement des larves d'*Acanthoscelides obtectus* Say (coléoptère Bruchidae). Ent. Exp. & App. 28/I (in press).

Authors' addresses:
Laboratory of Applied Zoology & Parasitology
Faculty of Agriculture
Thessaloniki
Greece

I.B.E.A.S. (ERA CNRS 328)
Université F. Rabelais
Parc de Grandmont
37200 Tours
France

POLYMORPHISM AND ECOLOGICAL REACTIONS IN *CALLOSOBRUCHUS MACULATUS* F. (COLEOPTERA, BRUCHIDAE) IN UPPER VOLTA

A.P. OUEDRAOGO & J. HUIGNARD
(Ouagadougou, Upper Volta/Tours, France)

INTRODUCTION

The larvae of *Callosobruchus maculatus*, the southern cowpea weevil develop within the beans of cowpea (*Vigna unguiculata*), making them inedible. Thus, the major part of the cowpea crops stored in traditional granaries in the Sahel zone of Africa is damaged by *Callosobruchus maculatus* and another Bruchidae, *Bruchidius atrolineatus.* Such extensive damage checks the development of cowpea crops; the fact is the more serious as this plant is a staple food in the native diet throughout Western Africa. The beans have a high protein content, with most of the amino-acids essential for growth (Smartt, 1976); the fresh leaves are eaten in many different dishes; after the pods are picked, the dried leaves are used for livestock feed during the dry season. Increasing cowpea production then becomes a necessity if one is to supply the local population with a sufficient protein intake. This early-domesticated plant, native to Western Africa (Steel *et al.*, 1980) is well adapted to the natural conditions prevailing in the Sahel.

Two varieties of *Vigna unguiculata* are grown in Upper Volta an early-flowering one which is sown as soon as the rainy season starts, and a late one. The farmers do so to secure at least one crop, even when the rainy season is short, as often happens in the Sahel zone.

Callosobruchus maculatus, as well as other Bruchidae like *Acanthoscelides obtectus* (Larson *et al.*, 1938; Hoffmann *et al.*, 1962), can reproduce either in the fields prior to the harvest and in the stored beans. In the fields *C. maculatus* females lay their eggs on dry pods (Alzouma, 1981). The eggs are scattered all over the pod; the first instar larvae bore into the pericarp, then into the seed coats before reaching the cotyledons. As contaminated beans are harvested, the emergent adults breed in the stores kept in granaries in which, in appropriate temperature conditions, generation may follow generation.

Pest control has been carried out in the stores and some traditional techniques (e.g. storage in ashes) occasionally achieve fair results. But we think that intervention in the field to stop oviposition on the dry pods is necessary for efficient protection of the crops. This entails some thorough knowledge of this insects life cycle and behaviour in this Sahel area in which a November to May dry season alternates with a May to October rainy one.

Callosobruchus maculatus, which may originate in Western Africa (Southgate, 1978) is known in two imaginal forms differing, in morphology, physiology and behaviour (Utida, 1954–1972):

– the flight form, which is the field form (Taylor *et al.*, 1974), is distinguished in the laboratory by its greater longevity and low fecundity;

– the flightless form, found in stores (Utida, 1954) is unable to fly; bred in the laboratory in the same conditions as the flight form, it is distinguished by a shorter imaginal life and a higher fecundity.

Utida (1972) finds that this phase polymorphism may be caused by the sensitiveness of certain genotypes to such ecological factors as crowding.

We feel the need, though, for taking up this problem by making allowances for the climatic conditions in these insects' natural environments, particularly during the dry season.

The present paper will move from a survey of the extensive damage caused in Upper Volta fields, mostly by both imaginal forms of *Callosobruchus maculatus*, onto a laboratory study of the conditions attending the reproduction of both imaginal forms of *Callosobruchus maculatus* from the infested crops.

STUDIES OF BRUCHIDAE EGGS DEPOSITED ON PODS OF VIGNA UNGUICULATA HARVESTED ON FIELDS NEAR OUAGADOUGOU – THEIR OFFSPRING

32 sets of 50 pods each were taken from cowpea fields at the end of the rainy season, from 8 to 18 October 1979, within an area 15 km east of Ouagadougou. There are no human dwellings in the vicinity of the sampled fields. Three sets of pods from the centre and the outlying parts of each field were brought to the laboratory. The eggs sticking to the pods, then the adults emerging from the beans, were counted.

The eggs on the exclusively indehiscent pods are usually of two types (Alzouma, 1981):

– *Bruchidus atrolineatus* eggs sticking along the suture;

– *Callosobruchus maculatus* eggs scattered all over the pericarp.

The following data were determined with respect to each set of pods:

– number of eggs of either species found on the pods;

– number of adults emerging from the seeds;

– concerning *Callosobruchus maculatus*, percentage of emerging adults of either flight or flightless forms.

Results (Table 1)

26% of the 1600 pods harvested at that time in the Ouagadougou area were infested with eggs of *C. maculatus* and *B. atrolineatus.*

78.5% of the 568 counted eggs had been laid by *B. atrolineatus* females, 21.5% by *C. maculatus* females.

Both species have a high post-embryonary death rate since 59% of *B.*

Table 1. Number of eggs deposited on pods collected in fields near Ouagadougou in October 1979. Study of their offspring.

			B. atrolineatus		*C. maculatus*		
Pods	Pods with eggs	Nb. of eggs	Eggs	Adults	Eggs	Flight form	Flightless form
1600	368	568	446	264	122	42	14

atrolineatus and 54% of *C. maculatus* eggs deposited on the pods have any progeny.

25% of the 56 observed *C. maculatus* adults belonged to the flightless form. It is interesting to note that they might have emerged in nature if the pods had not been harvested. Alzouma (1980) also records the emergence of flightless adults from wild *Vigna* pods found in Niger fields during the dry season. One may well wonder what happens to these adults of poor migratory ability when they emerge in nature.

The substantial death-rate during post-embryonary development in both species may be caused by the way the deposited eggs are located on the pods, For instance, the eggs sticking on the pericarp between two beans (Fig. 1) do stand a chance of evolving into adults (Huignard, unpublished). The first instar larva will bore into the pericarp, fall inside the pod and die. The reason is that, unlike those of *Acanthoscelides obtectus* (Labeyrie, 1962), the first instar larvae of *C. maculatus* or *B. atrolineatus* cannot move about since their legs are non-functional. Besides, the death-rate is substantially lower when the eggs are deposited right on the beans (Ouedraogo, 1978). For all that, other factors such as climatic conditions, pod texture, hardness of the bean tegument, may also account for it.

LABORATORY STUDY OF THE EXTERNAL FACTORS (HOST-PLANT-MATING) INFLUENCING OOGENESIS AND OVIPOSITION IN BOTH IMAGINAL FORMS OF CALLOSOBRUCHUS MACULATUS

The flightless form

Methods. The strain of *C. maculatus* used in the experiment originated from Bobo Dioulasso; it was reared for several generations on beans of *Vigna unguiculata* in a climatic box (29 °C ± 2 Constant – 70% H.R. – absolute darkness).

After several generations had been reared under these conditions, the percentage of the emergent flight forms was very low (Ouedraogo, 1978).

The females separated after emergence were put in the following conditions:

Experiment 1: females kept virgin and isolated.

Experiment 2: virgin females in the presence of seeds of *Vigna unguiculata.*

Experiment 3: females mated (a few hours after emergence) and isolated.

Experiment 4: females inseminated and put with cowpea seeds.

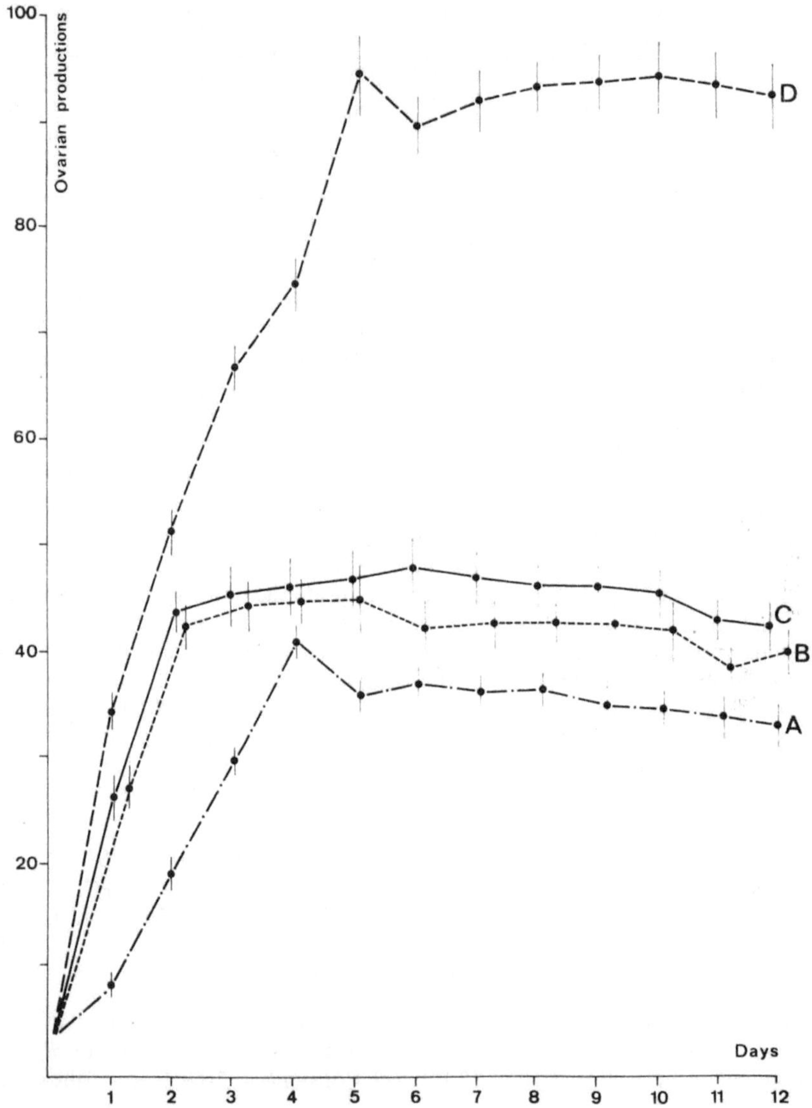

Fig. 1. Evolution of ovarian production during experiment A: virgin females: B: virgin females in presence of seeds; C: mated females; D: mated females in presence of seeds.

In every case the experiment lasted 15 days, during which period the insects were fed some regularly renewed honeyed water.

Each set included 240 females at first; 20 females were studied daily in order to determine:

– their fecundity (number of eggs laid);

– their ovarian production (number of eggs laid + number of ripe chrionated ovocytes in retention in lateral oviducts).

Results (Fig. 1). – Virgin females kept isolated lay no eggs. However, oogenesis starts in the first few days of imaginal life, and ripe chorionated oocytes accumulate in the lateral oviducts. The number of oocytes in retention gradually increases until the 4th or 5th day of imaginal life, then becomes stable (Fig. 1). Oogenesis stops as was shown in virgin females of *Acanthoscelides obtectus* (Huignard, 1975; Biemont, 1979).

Introducing *Vigna* seeds stimulates oogenesis without inducing oviposition. This is made clear by the fact that ovarian production increases faster in that case than in virgin females during the first few days of imaginal life. Ovarian production subsequently becomes stable with recorded values always above those of the virgin females.

Mating results in a spermatophore being emitted into the females' bursa copulatrix (Ouedraogo, 1978). This spermatophore, elaborated during copulation by the male's five pairs of accessory glands, contains spermatozoa that rapidly migrate along the spermatophore membrane towards the spermatheca (Fig. 2A and B).

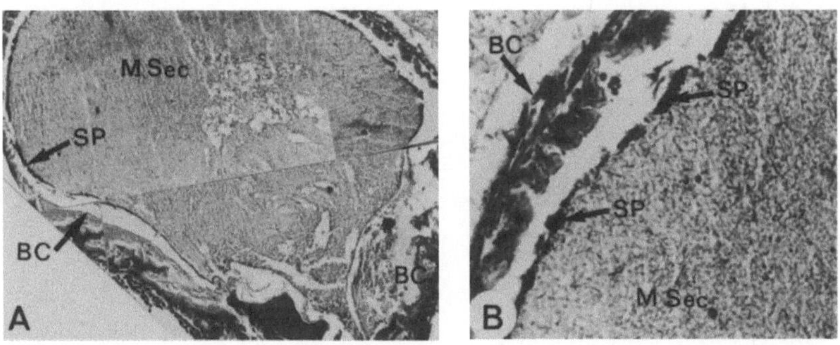

Fig. 2. Longitudinal section of a spermatophore in the bursa copulatrix (Bouin Masson Trichrome). bc: bursa copulatrix; m sec: male secretions; sp: spermatozoa. A(× 63) B(× 126).

While mating stimulates oogenesis, it induces oviposition in but a few females (of very low fecundity at that). This stimulating effect becomes apparent in the first few days of imaginal life as ovarian production increases more rapidly than in the reference virgin females. Oogenesis stops subsequently, and ovarian production is more or less identical to that found in females put on *Vigna* beans.

The factor that causes the most substantial stimulation of oogenesis and induces oviposition is the interaction between *Vigna* seeds and mating, which was observed in other Bruchidae too (Huignard, 1975; Pierre, 1980). Ovarian production is seen to increase rapidly during the first few days of imaginal life, then it becomes stable, with a mean value set between 90 and 100.

Observation (Fig. 3) of the day to day evolution of fecundity brings to notice the fact that oviposition rises to a maximum within 24 or 48 h of mating and gradually decreases afterwards. The resulting oviposition curve is identical to that found in the other Bruchidae similarly investigated (Huignard, 1975).

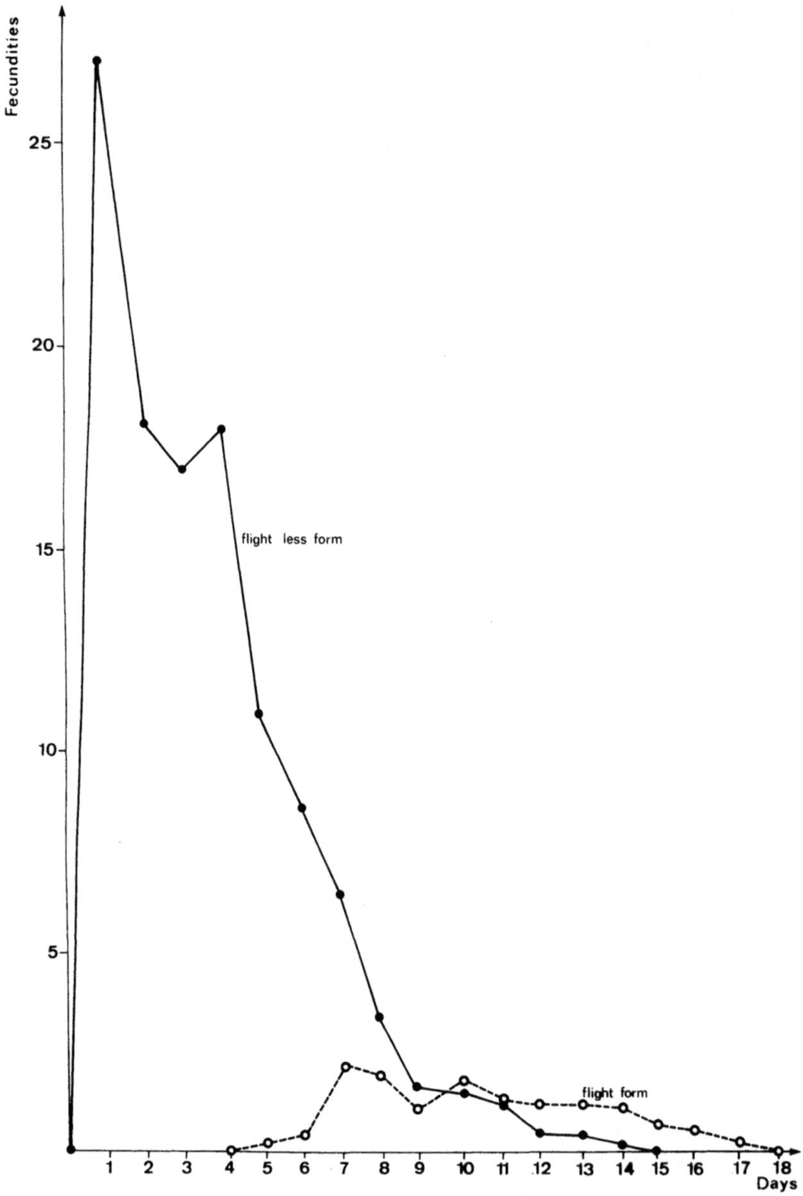

Fig. 3. Evolution of fecundities of inseminated females in presence of seeds (flightless and flight form).

The flight form

The females under study originated from eggs gathered in nature on ripe pods at Bobo Dioulasso. They were investigated under the same temperature, photoperiodical and hygrometric conditions as were the flightless adults.

Virgin females reared without seeds of Vigna unguiculata (Table 2). 80 virgin females were isolated as stated above from the start of imaginal life. During the 30 day experiment, a number of females were dissected at various periods. None of them, whatever the age, was then found to have formed a single ripe chorionated oocyte. The ovarioles only contained previtellogenetic oocytes (Fig. 4).

Table 2. Reproductive activity of flight form females after a 20 days experiment.

	Number of females	Fecundities	Ovarian productions
Virgin females	80	0	0
Virgin females with cowpea seeds	80	0	12.7 ± 3.2
Inseminated females with cowpea seeds	25	13.7 ± 3.4	25.4 ± 3.2

This evidence shows that under experimental conditions in the laboratory the ovarian activity of flight females is nil. They are similar to the females of *Acanthoscelides obtectus* found in some Colombian ecosystems; when the latter are kept virgin, their ovarian activity is also nil (Huignard *et al.*, 1978, 1979).

Virgin females isolated without seeds of Vigna unguiculata. The presence of *Vigna* beans stimulated the females' ovarian activity. The ovarioles present vitellogenetic phases and ripe chorionated oocytes accumulated in the lateral oviducts (Table 2) but no oviposition ever occurred throughout the experiment.

Inseminated females in presence of Vigna unguiculata seeds. 25 females were put from emergence in the presence of males in Petri dishes containing cowpea seeds. It was found that the females did not mate in the first few days of imaginal life. It is possible, as was shown in other insects (Adams *et al.*, 1969; Huignard *et al.*, 1979), that the females' sexual receptivity is dependent on their ovarian condition. This can be inferred from the fact that the females mated and started oviposition after a number of ripe oocytes had been formed under the stimulating influence of the beans. Still, under experimental conditions, very few eggs were ever laid, and a number of oocytes remained in retention in the lateral oviducts (Table 2).

This low reproductive activity of the flight females has already been reported by Utida (1972), but it may be linked to the experimental conditions (constant temperature, confinement, no pods), and further investigation is necessary.

DISCUSSION

Observations based on cowpea pods taken from fields near Ouagadougou at the end of the rainy season show the economic significance of *Bruchidus*

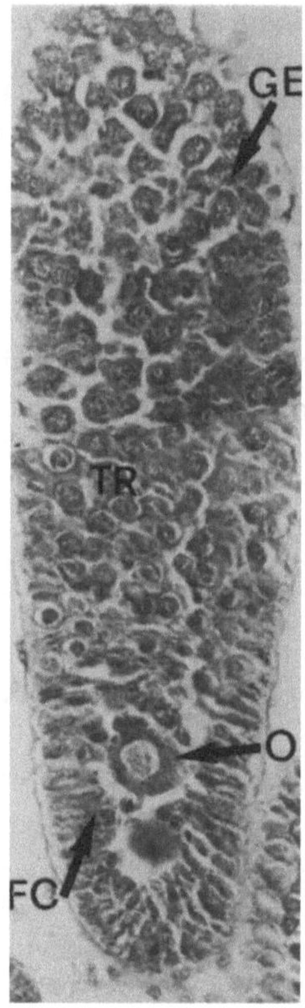

Fig. 4. Longitudinal section of ovarioles of a virgin flight form females 10 days old. (Bouin Masson trichrome – × 180). ge = germarium; oo = oocytes in previtellogenesis; tr = trophocytes; cf = follicular cells.

atrolineatus and *Callosobruchus maculatus* in the Sahel. Although 26% only of the pods are infested with eggs, that is enough to contaminate the bean harvest to a degree. Contamination will increase during storage along with the passing of generations. This proves the necessity of protecting cowpea plantations before the dry pods are harvested in order to prevent those two Bruchidae from multiplying in the stored seeds. As a result, one is faced, concerning *Callosobruchus maculatus*, with the same problem as *Acanthoscelides obtectus* poses (Labeyrie, 1981) and, here again, precise ecological studies must be undertaken.

Callosobruchus maculatus eggs laid on pods evolve into both of the imaginal forms described by Utida (1954–1972), with greater frequency for the flight form. However, a number of flightless adults do emerge from the pods; they would exist in nature during the dry season had these beans not been harvested. Thus, it can be assumed that this rather inactive stage may be a phase that enables a few individuals to survive in shelters (rolled-up leaves, tree barks, as was found with various Bruchidae (Biemont *et al.*, 1980; Genduso, 1978).

We suggest that this flight form may have adapted to storage conditions following man's domestication of this plant. When man brings to the store cowpea seeds contaminated by *C. maculatus*, those individuals that breed and gradually become dominant there with each successive generation are mostly non-migratory flightless ones. That selective effect is the stronger as flight females have a low fecundity on stored cowpea seeds.

Granting this, it is sheer nonsense to refer to the flight form as the nature form and the flightless form as the one found in stores.

We think it necessary to take up these studies both in the field and in the traditional African granary in order to analyse the significance of these two forms and their importance for the survival of populations in the dry season and at the start of the rainy season when ripe cowpea pods are not available.

The reproduction of flight females is strictly controlled by external factors; they form oocytes only in the presence of cowpea seeds and vitellogenesis must start before they accept the males and mate. It follows that reproduction can only take place when females are in the presence of the substratum of oviposition (ripe pods or seeds). This avoids a lengthy retention of ripe oocytes in the lateral oviducts since it might impair their capacity for later development, as shown in *Acanthoscelides obtectus* (Biemont, 1979). The reproduction of flightless females is to some extent independent of external factors (*Vigna* beans, mating) since isolated virgin females produce some thirty ripe oocytes accumulating in the lateral oviducts. However, the insects used in these experiments originated from strains reared on beans in the laboratory over several generations, that is, under conditions consistently favourable to reproduction and the developement of progeny. In this case, selective pressures may be modified so that the reproductive function is less strictly regulated, as was observed in some strains of *Acanthoscelides obtectus* found in nature and continuously reared on beans in the laboratory (Huignard *et al.*, 1981).

Further study may be necessary, concerning the ovarian activity of flightless females as they emerge from ripe pods at the start of the dry season, in order to examine how their reproduction is regulated. This investigation will help to build up a clear picture of the significance of those two phases in *Callosobruchus maculatus*, and of their importance in the life cycle of this insect developing in such an arid tropical zone as the Sahel; without that information no efficient pest control can be suggested.

REFERENCES

Adams, T.S. & Hintz, A.M., 1969. Relationship of age, ovarian development and the corpus allatum to mating in the house fly *Musca domestica*. J. Insect Physiol. 15: 201–215.

Alzouma, I., 1981. Observations on the ecology of *Bruchidius atrolineatus* and *Callosobruchus maculatus* in Niger, this volume.

Biemont, J.C., 1979. Influence de la plante hôte et de la copulation sur la levée de l'inhibition du développement ovarien liée à la rétention des ovocytes chez *Acanthoscelides obtectus*. Ann. Soc. Entomol. Fr. (NS) 15: 93–99.

Biemont, J.C. & Bonet, A., 1981. Preliminary observations of bean weevil populations from the Acanthoscelides obtectus Say. group living on wild or subspontaneous

Phaseolus vulgaris L. and Phaseolus coccineus L. and on Phaseolus Vulgaris L. cultivated in the Tepoztlan region state of Morelos, Mexico, this volume.
Genduso, P., 1978. Insectes nuisibles aux légumineuses en Sicile et observations sur l'hivernage des Bruchides univoltines. Boll. 1st Ent. Agr. Oss. Fitopato Palermo 10: 169–176.
Hoffmann, A., Labeyrie, V. & Balachowsky, A.S., 1962. Famille des Bruchidae in Entomologie appliquée à l'agriculture. A.S. Balachowsky. Masson, Paris, 564 pp.
Huignard, J., 1975. Influence de la copulation sur la fonction reproductrice femelle chez Acanthoscelides obtectus: II Analyse expérimentale. Ann. Sci. Nat. Zool. 12–17: 1–66.
Huignard, J. & Biemont, J.C., 1978. Comparison of four populations of *Acanthoscelides obtectus* (Coleoptera Bruchidae) from different Colombian ecosystems. Assay of interpretation. Oecologia 35: 307–318.
Huignard, J. & Biemont, J.C., 1979. Vitellogenesis in *Acanthoscelides obtectus* (Coleoptera Bruchidae). II: The conditions of vitellogenesis in a strain from Colombia. Comparative study and adaptative significance. Int. J. Invert. Reprod. 1: 233–244.
Huignard, J. & Biemont, J.C., 1981. Reproductive Polymorphism of populations of *Acanthoscelides obtectus* from different Colombian ecosystems, this volume.
Labeyrie, V., 1962. Les *Acanthoscelides* in Entomologie appliquée à l'agriculture A.S. Balachowsky 469–484: Masson, Paris, 564 pp.
Labeyrie, V., 1981. Ecological problems arising from weevil infestation of food legumes, this volume.
Larson, A.O. & Fischer, C.K., 1938. The bean weevil and Southern cowpea in California. U.S.D.A. Tech. Bull. 593: 1–71.
Ouedraogo, P.A. Etude de quelques aspects de la biologie de *Callosobruchus maculatus* et de l'influence des facteurs externes stimulants (plante hôte et copulation) sur l'activité reproductrice de la femelle. Thèse 3ème Cycle, Université de Toulouse (France), 101 pp.
Pierre, D., 1980. Influence des graines ou des gousses Mûres de la plante hôte (*Phaseolus vulgaris*) sur l'activité reproductrice de *Zabrotes subfasciatus* (Coleoptera Bruchidae). C.R. Acad. Sci. Paris 290: 1007–1010.
Smartt, J., 1976. Tropical pulses. Longman, 348 pp.
Southgate, B.J., 1978. The importance of the Bruchidae as pests of grain legumes, their distribution and control. In: Pests of Grain Legumes: Ecology and Control Academic Press. 219–229.
Steele, W.M. & Mehra, K.L., 1980. Structure, evolution and adaptation to farming systems and environments in *Vigna*. In: 'Advances in Legume Science'. 393–404, Royal Botanic Garden, Kew, 667 pp.
Taylor, T.A. & Agbaje, L.A., 1974. Flight activity in normal and active forms of *Callosobruchus maculatus* in a store in Nigeria. J. Stored Prod. Res. 10: 9–16.
Utida, S., 1954. Phase dimorphism observed in laboratory population of the cowpea weevil *Callosobruchus maculatus.* Jap. J. appl. Zool. 18: 161–168.
Utida, S., 1972. Density dependent polymorphism in the adult of *Callosobruchus maculatus.* J. Stored Prod. Res. 8: 111–125.

Authors' addresses:
Institute Supérieur Polytechnique
B.P. 7021
Ouagadougou
Haute-Volta

I.B.E.A.S. (ERA CNRS 328)
Université F. Rabelais
Parc de Grandmont
37200 Tours
France

THE USE OF RESISTANT VARIETIES OF COWPEAS (*VIGNA UNGUICULATA*) TO REDUCE LOSSES DUE TO POST-HARVEST ATTACK BY *CALLOSOBRUCHUS MACULATUS*

P. DOBIE
(*Slough, Berks, U.K.*)

INTRODUCTION

Cowpeas (*Vigna unguiculata*) are an important staple food crop over much of West and Central Africa, and are also grown in many other parts of the world. The full yield potential of cowpeas is seldom realized due to the interaction of many climatic, ecological and edaphic factors, and loss caused by pre-harvest insect infestation is one of the most important of these factors. In Nigeria, Raheja (1976) showed that pre-harvest pests could reduce the yield of cowpeas by 92–97% of the potential yield in the absence of insects. In order to obtain satisfactory yields of cowpeas it is therefore necessary to spray the crop several times during its growth which adds greatly to the cost of production. Cowpeas are therefore a valuable and expensive crop, and the further extensive losses which are caused by post-harvest infestation by *Callosobruchus maculatus* are therefore of great concern to those involved in storing or marketing the crop. None of the available estimates of the actual quantity of legumes lost as a result of bruchid attack have been obtained using acceptable and reliable loss assesment techniques (Harris & Lindblad 1978; Anon 1978). However the estimates which do exist indicate that appreciable quantities are lost to bruchids every year. Caswell (1973) estimated on the basis of samples collected in Nigerian markets that about 55,000 tons of cowpeas out of a 1 million ton harvest were consumed yearly by Bruchidae, although he regarded this as a conservative estimate. This estimate did not take into account any cowpeas which were rejected as damaged before sale by the producer or market stallholder, and did not estimate the loss of cowpeas kept by the producer for consumption by his family. Staff at the Tropical Products Institute frequently find during their work with subsistence farmers that legumes are not stored in quantity on the farm due to their high susceptibility to insect attack. Frequently they are sold immediately after harvest to raise cash only to be bought back at a high price when required. In North-East Brasil damage caused by *C. maculatus* is recognized as the major constraint against increasing cowpea production (E. Watt, personal communication).

It can be fairly concluded that the effects of post-harvest attack by

Callosobruchus spp. are severe, and that reduction of the level of attack should form an integral part of any strategy to improve cowpea production.

INTERNATIONAL COWPEA PRODUCTION

The Grain Legume Improvement Program of the International Institute of Tropical Agriculture (IITA) in Ibadan, Nigeria is involved in improving cowpea production. Their mandate is global: materials are being developed at IITA for growth in a wide range of environments in all cowpea-growing countries. IITA have a large and active breeding programme, within which considerable effort is being expended towards improving the resistance of cowpea lines to insect pests. As part of this effort IITA sought the assistance of the Tropical Products Institute (TPI) of the Overseas Development Administration of the United Kingdom to assist in the development of cowpea varieties which are resistant to attack by *Callosobruchus maculatus.* This paper outlines the developments which have been made in this collaborative research, and discusses the potential benefits of this approach to post-harvest loss reduction.

THE DISCOVERY OF A RESISTANT VARIETY

Early work at both IITA and TPI showed that nearly all cowpea varieties were highly susceptible to attack by *C. maculatus* and that there was very little variability among these varieties in susceptibility. However Dr. S.R. Singh at IITA screened about 5,000 varieties from the IITA bank of field-collected cowpeas and discovered one variety from Northern Nigeria to be highly resistant to *C. maculatus* attack. Unfortunately this variety is highly susceptible to virus attack, which made seed multiplication difficult. Eventually, however, a quantity of seed was made available to entomologists outside Nigeria, and a sample was sent to TPI for further study. The variety is known at TVu 2027.

INVESTIGATION OF THE RESISTANCE Of COWPEA VARIETY TVu 2027

Investigations at TPI quickly confirmed that TVu 2027 was highly resistant to a strain of *Callosobruchus maculatus* which had originated in Campinas, Brasil. Although *C. maculatus* adults readily laid eggs upon the surface of the seeds and the larvae successfully hatched from the eggs and began feeding on the underlying cotyledons, growth of the larvae was extremely slow, and eventually most or all of the larvae died. In control susceptible varieties, between 80 and 95% of hatched larvae developed to adults (Gatehouse *et al.* 1979). This strongly suggested the presence of a toxic compound in the cowpea cotyledon, so TPI provided a grant to permit Dr. A Gatehouse at the University of Durham, England, to study the secondary chemical compounds of the seeds of TVu 2027.

It has been postulated by several authors that one function of certain secondary chemicals in legume seeds is to provide protection against insect attack, although to our knowledge the theory had never been proved in any study of the effect of a secondary compound extracted from a seed upon the pests of that seed.

In her study of secondary compounds, Dr. Gatehouse was unable to detect saponins or lectins in the seed of TVu 2027, and preliminary experiments kindly carried out in the laboratory of Dr. A. Bell, Kings College, London, showed no significant differences in the type or quantity of the major non-protein amino acids in TVu 2027 and control varieties. Dr. Gatehouse, however, showed that TVu 2027 contained a trypsin inhibitor which was present at about twice the level of that of the highest susceptible control (Gatehouse *et al.* 1979). Trypsin inhibitors have been implicated as anti-metabolic substances in legumes (e.g. Applebaum *et al.* 1974), so further studies were carried out to investigate the effect of the inhibitor. A quantity of the inhibitor was extracted from cowpea seeds and purified. The purified inhibitor was then mixed with the ground cotyledon of cowpea variety TVu 57 at the levels of 0.1%, 0.5% and 0.8%. The TVu 57 contained 0.25% inhibitor, while the TVu 2027 contained 0.92% inhibitor.

It was known that *C. maculatus* adults will lay eggs on the taut surface of a thin, flexible film used for wrapping food, and that the hatched larvae can penetrate the film and survive to the adult stage if a suitable food is provided beneath the film. This method was used to determine whether *C. maculatus* larvae could survive on the treatments listed above. For each treatment, ten shallow plastic dishes (17 mm dia. × 3 mm deep, formed by modifying the lids of glass vials obtained from Scientific Supplies Co. Ltd., London) were filled with the appropriate meal or mixture and then wrapped in the plastic film. The film used was 'Cling n' Seal' (W.H. Smiths, UK), a PVC film of about 15 μm thickness. The prepared dishes were placed in a jar of cowpeas infested by *C. maculatus*, and female insects laid eggs on the surface of the film. After 24 h the dishes were removed and placed individually in 5 × 2.5 cm glass tubes closed by plastic foam stoppers. After one week the number of hatched eggs on the film in contact with the meal on each dish was counted. Twenty days after oviposition the dishes were opened and examined, and the numbers of surviving larvae not passing through a 500 μm mesh were counted. (These experiments were carried out at 27 ± 1 °C and 70 ± 5% rh).

The percentage survivals of eggs laid on each treatment are given in Table 1. Although there was some survival on the TVu 2027, survival became reduced with an increased level of trypsin inhibitor in the diet. A level of 1.05% total inhibitor killed all developing larvae.

THE INHERITANCE OF RESISTANCE

In order to determine how high levels of trypsin inhibitor and resistance to *C. maculatus* are inherited in cowpeas, a series of crossing trials were started at IITA.

Table 1. The survival of *C. maculatus* larvae developing on diets incorporating different levels of cowpea trypsin inhibitor.

Cowpea var. forming base of diet*	Assayed inhibitor content (%)	Added inhibitor (%)	Estimated inhibitor content of diet (%)	Eggs laid	% survival
TVu 57	0.25	0.0	0.25	72	78.1
TVu 57	0.25	0.1	0.35	110	80.1
TVu 57	0.25	0.5	0.75	120	20.0
TVu 57	0.25	0.8	1.05	74	0.0
TVu 2027	0.92	0.0	0.92	47	32.2

* the base of each diet consisted of ground cotyledon of the indicated cowpea variety.

Hand pollinations were carried out between TVu 2027 and several virus resistant varieties. The results given here concern crosses between TVu 2027 and TVx 1193–9F.

Seed of plants produced by making the following crosses was sent to TPI for bio-assay and to the University of Durham for chemical assay. (In describing all crosses, the female parent will be listed first).

Parents

TVu 2027
TVx 1193–9F

F_1 *crosses*

TVx 1193–9F x TVu 2027
TVu 2027 x TVx 1193–9F

Backcrosses

(TVx 1193–9F x TVu2027) x TVx 1193–9F
TVx 1193–9F x (TVx 1193–9F x TVu 2027)
(TVx 1193–9F x TVu 2027) x TVu 2027
TVu 2027 x (TVx 1193–9F x TVu 2027)

F_3 *cross*

TVx 1193–9F x TVu 2027

For the parent, F_1 and backcross seed, up to 25 seeds of each material were placed individually in 5 x 2.5 cm glass tubes in random order. A single female *C. maculatus* adult aged between 0 and 48 h was introduced into each tube and left to lay eggs for 4 h. One week later the seeds were examined, and any pea with fewer than 4 eggs laid on the surface was exposed to oviposition for a further 4 h. All eggs were counted, and successfully emerging adults were subsequently removed and counted during the emergence of the F_1 insects. The results are given in Figs. 1–3. Each histogram shows the frequency of occurrence of peas with given percentage survivals of developing

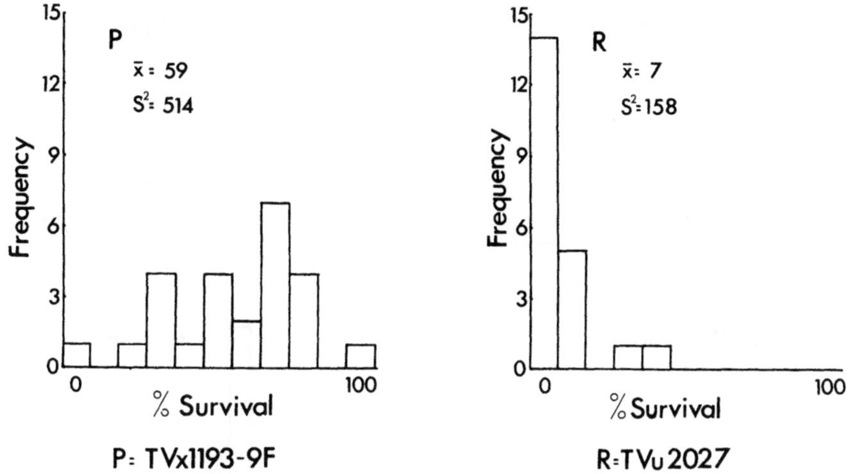

Fig. 1. The survival of *C. maculatus* larvae developing on cowpea varieties TVx 1193–9F and TVu 2027. The histograms show the number of seeds (frequency) with indicated levels of larval survival (grouped into 10% intervals).

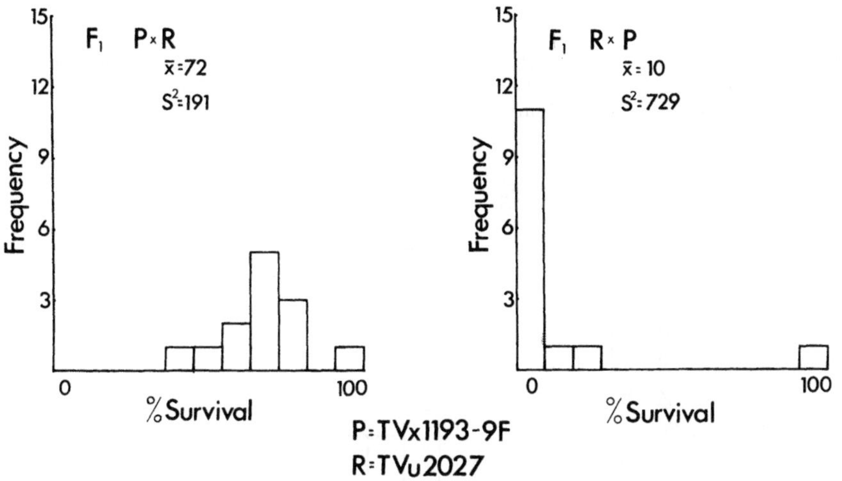

Fig. 2. The survival of *C. maculatus* larvae developing on F_1 crosses between cowpea varieties TVx 1193–9F and TVu 2027. The histograms show the number of seeds (frequency) with indicated levels of larval survival (grouped into intervals of 10%).

larvae. (The data are grouped into intervals of 10%). Fig. 1 contrasts the survival patterns on TVx 1193–9F and TVu 2027. Fig. 2 demonstrates a large maternal effect in inheritance within the F_1 crosses where the resistant seed comes from a cross in which the female parent was resistant. This

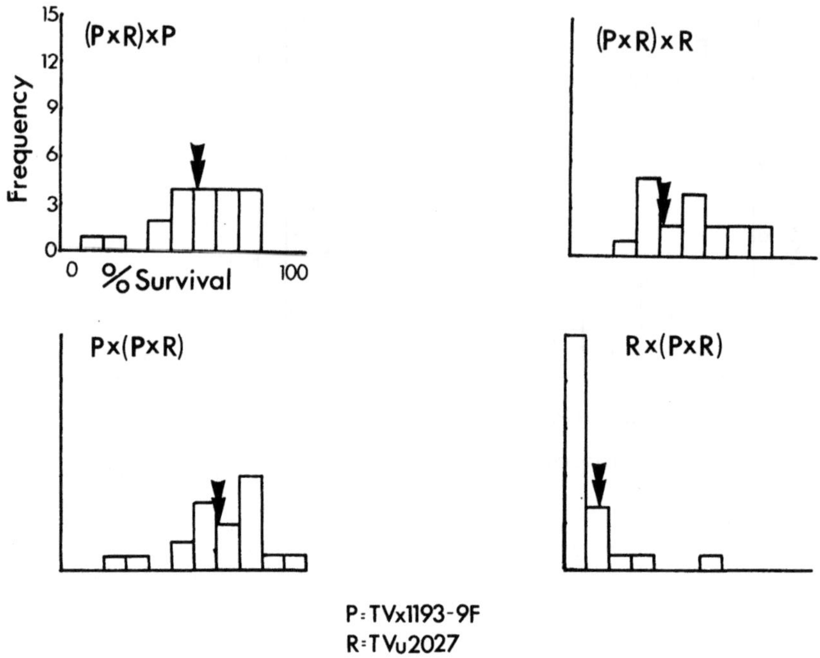

Fig. 3. The survival of *C. maculatus* larvae developing on backcrosses of cowpea varieties TVx 1193–9F and TVu 2027. The histograms show the number of seeds (frequency) with indicated levels of larval survival (grouped into 10% intervals).

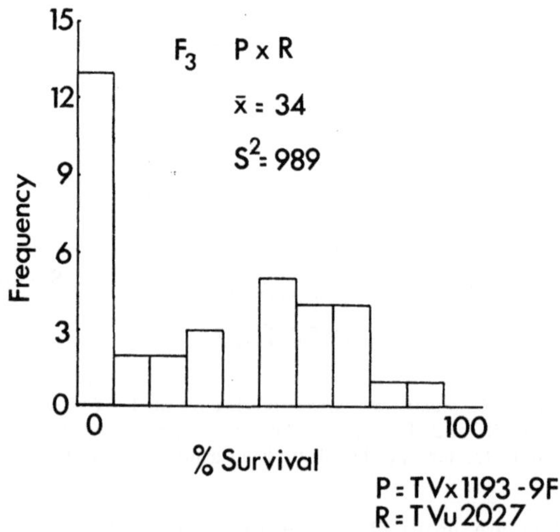

Fig. 4. The survival of *C. maculatus* larvae on seed of the F_3 cross between cowpea varieties TVx 1193–9F and TVu 2027. The histogram shows the number of plants (frequency) with indicated mean levels of larval survival (grouped in 10% intervals).

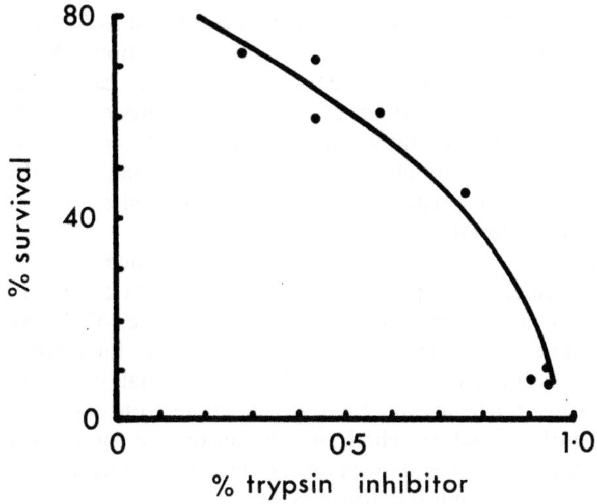

Fig. 5. The relationship between % survival of *C. maculatus* and the concentration of trypsin inhibitor.

effect is further demonstrated for the backcrosses in Fig. 3. Only the cross involving a resistant female parent produced resistant seed.

Similar assays were carried out on the seed of the F_3 plants. In this case, however, seed from different plants was available. Fig. 4 shows the pattern of distribution of the mean percentage survival of eggs laid on 25 seeds each of 35 plants (i.e. the histogram shows plant to plant differences rather than seed to seed variation). Analysis of variance showed that the plants with between 0 and 10% survival to be significantly different from the others.

For seed of all of the crosses listed above, trypsin inhibitor assays were carried out. Fig. 5 shows the relationship between trypsin inhibitor content and mean percentage survival for each cross.

CONCLUSIONS AND DISCUSSION

From the results reported in this paper, the following conclusions can be drawn:

1 Most cowpea varieties are highly susceptible to attack by *C. maculatus*, but a single variety TVu 2027 has been shown to be resistant.

2 Resistance is associated with high levels of trypsin inhibitor.

3 Resistance of this type is inheritable, with a maternal effect being shown.

Further research is currently underway to investigate the stability of this resistance. There have been some reports of strains of *C. maculatus* which will attack TVu 2027 (Huignard, personal communication). However it appears possible that TVu 2027 from the IITA collection is not a pure variety, and that there is some variability in resistance to *C. maculatus* among different samples (R. Redden personal communication). Nevertheless, the possibility that the release of resistant varieties may cause the selection of biotypes of *C. maculatus* which are able to feed upon TVu 2027 or other varieties with high trypsin inhibitor contents must be considered. In order to anticipate such an eventuality research is continuing into the search for alternative sources of resistance. Resistance to oviposition and the resistance of seed pods to penetration are both being considered.

The discovery of an effective source of resistance in a crop which is generally extremely susceptible to post-harvest attack should encourage a further search for alternative sources of resistance. The development of resistant varieties will enable advances to be made in the reduction of losses, especially at subsistance farmer level where capital for insect control is usually lacking. Such an approach to loss reduction does not require the introduction of technology which may be unsuitable for small scale farmers. The widespread use of resistant varieties may also reduce the demands made upon conventional chemical control techniques.

REFERENCES

Anon, 1978. Post-harvest food losses in developing countries. National Academy of Sciences, Washington, vii + 206 pp.

Applebaum, S.W., Birk, Y., Harpaz, I. & Bondi, A., 1974. Comparative studies on proteolytic enzymes of Tenebrio molitor. Comp. Biochem. Physiol. 11 (1): 85–103.

Caswell, G.H., 1973. The impact of infestation on commodities. Trop. stored Prod. Inf. 25: 19.

Gatehouse, A.M.R., Gatehouse, J.A., Dobie, P., Kilminster, A.M. & Boulter, D. 1979. Biochemical basis of insect resistance in Vigna unguiculata. J. Sci. Food Agric. 30: 948–958.

Harris, K.L. & Lindblad, C.J. 1978. Post-harvest grain loss assessment methods. Am. Assoc. Cereal Chem. xiii + 193 pp.

Author's address:
Tropical Products Institute
London Road
Slough
Berks. United Kingdom

BRUCHIDAE RELATED TO GRAIN LEGUMES IN THE AFRO-TROPICAL AREA

J. DECELLE
(*Tervuren, Belgium*)

The taxonomy of Bruchidae has long been neglected. Establishing the precise identity of species of economic interest has been and still is often a problem for the non-specialist entomologist. The determiner's work is made more complex by the vagueness of description of known species, the frequent lack of evident external characters, the uniform shape and colour of many species, the polymorphism of some of them, and the easily introduced exotic forms. It means that bibliographical data, mainly the early ones, are sometimes to be considered reservedly as far as the exact identity of the mentioned or studied Bruchidae is concerned.

In the course of works upon the taxonomy of African Bruchidae, we have examined important materials kept in the collections of many scientific institutions, and thus we have gathered documents on the geographical distribution of species, and data, which are unfortunately more scanty, on their host plants. Taking into account these documents and the critical examination of the bibliography, fifteen species detrimental to grain legumes can at present be mentioned in the Afrotropical area. Either they are originating from that area, or they have colonized it after being passively introduced by human activity. Therefore, the species whose introduction was not followed by colonization will not be mentioned.

1. SUBFAMILY *BRUCHINAE*

A. Genus *Bruchus* L.

This palearctic genus comprises a certain number of species harmful to cultivated Fabaceae and mainly to Vicieae. Two of them have found favourable conditions of development in the most southern part of the African continent. *Bruchus* are univoltine species which oviposit on ripening pods, and are unable to reproduce in dry seeds.

1. Bruchus pisorum (L.). The pea-weevil, originating from the Middle-East like its host plant, has well settled in the South-West of the Cape province in South Africa, where it is considered as a pest.

Series Entomologica, Vol. 19: ed. by V. Labeyrie

It has been collected elsewhere, especially in Rhodesia, Tanzania, Madagascar and Mauritius, with imported seeds. It does not seem to have settled yet in the pea-crops at a high altitude, above 2000 metres, in various mountainous areas of East Africa.

2. Bruchus rufimanus Boh. Related to various *Vicia,* this species proves detrimental to broad bean crops (*Vicia faba*) in the Cape Province in South Africa. Incidentally collected elsewhere (Senegal, Tchad, Eritrea Saô Tome, Angola, Mozambique) it does not seem to have become indigenous in these areas.

B. Genus *Bruchidius* Schilsky

Bruchidius atrolineatus (Pic). Widely distributed in West, Central and East Africa, from Senegal to Eritrea in the north, from Angola to Mozambique in the south. Its most common host plant is *Vigna unguiculata,* better known by the name of niébé and cowpea. This bruchid has also been collected on *Vigna catjang* in Ghana and on *Phaseolus radiatus* in Tanzania.

It seems to be a univoltine species normally ovipositing on the ripening pods, and whose hatching occurs at harvest time or a little later; it does not seem to be able to reproduce in stored dry seeds.

Bruchidius atrolineatus certainly is of African origin. It has been found in Jamaica and in the State of Bahia in Brazil (where it has been described again by the name of *Bruchus semiflabellatus* Pic).

C. Genus *Callosobruchus* Pic

This genus with a paleotropical distribution comprises several species related to cultivated Fabaceae and mainly to Phaseolae. Some of them have now become subcosmopolitan in tropical and hot temperate areas and can subsist in colder areas under artificial conditions. Most species whose biology is known are polyvoltine and reproduce in dry seeds, and they can consequently prove very detrimental in stocks.

Six harmful *Callosobruchus* species are to be mentioned in Africa.

1. Callosobruchus analis (F.). This species very common in South and South-east Asia is probably originating from that area. In Africa, it has been found several times in the far eastern part of the continent (Sudan, Eritrea, Ehtiopia, Tanzania, Natal, South Africa), in Madagascar and Mauritius, more rarely elsewhere (Lower Zaïre, 1913). It seems to have rarely settled. In Sudan only it has been mentioned as a pest for dolichos (*Lablab purpureus*) and lentils (*Lens esculenta*).

2. Callosobruchus chinensis (L.). Originating from South-East Asia, *Callosobruchus chinensis* has been found in various African areas but it does not seem to have considerably developed as a pest, except in Madagascar and sometimes here and there. Other *Callosobruchus* (in particular *rhodesianus*) have been mistaken for *chinensis*.

Though it can generally exploit the seeds of many Phaseolae, Walker (1979) reports that in Swaziland, it only attacks *Phaseolus aureus*.

3. Callosobruchus maculatus (F.). It is the most widespread *Callosobruchus* and the most frequently collected in Africa. It is now very difficult to establish its precise origin, but it might well be African.

Its most frequently observed host plant is *Vigna unguiculata* in Africa, it has also been obtained from other Phaseoleae seeds (*Lablab purpureus* (dolichos), *Kerstingiella geocarpa, Voandzeia subterranea, Cajanus cajan, Phaseolus aureus* and *angularis.*

4. Callosobruchus phaseoli (Gyll.). Is found in various African regions: Guinea, Nigeria, Kenya, Tanzania, Uganda, Rwanda, Angola, South Africa (Natal, Transvaal), Madagascar, La Réunion. Its Asian or African origin cannot be precisely determined.

Its usual host-plant in Africa is *Lablab purpureus* (= *Dolichos*). In Angola, it has also been obtained from *Cicer arietinum.*

5. Callosobruchus rhodesianus Pic. This species which has often been mistaken for *C. chinensis,* is widely distributed over the greater part of Africa South of Sahara. There it is often collected outside the crops. While it exists in the Comoro Islands (Moheli), it has not yet been observed in Madagascar. The African origin of *C. rhodesianus* is undeniable.

That bruchid has been obtained from *Vigna unguiculata* and *Cajanus cajan* seeds. In Zaïre, it has been observed on *Phaseolus mungo* (probably = *Vigna campestris*?).

6. Callosobruchus subinnotatus (Pic). Localized in West Africa (Senegal, Mali, Ivory Coast, High Volta, Ghana, Togo, Nigeria, Cameroun, Tchad), this species was introduced in the Cape Verde Islands (Castel–Branco, 1965) and in Puerto-Rico (according to Southgate).

Its usual host plant is *Voandzeia subterranea,* a Phaseolinae with subterranean pods known by the names of voandzu and bambara nuts. Damage starts in the fields on the ripe pods near the surface of the soil; then it goes on after the harvest and in stores. In the Cape Verde Islands, has been mentioned from feijao-pedra (?) (Castel Branco, 1965).

D. Genus *Specularius* Bridwell

This genus now only comprises Afrotropical species whose known host plants are all Fabaceae Phaseoleae. Their biology has been poorly documented as yet. Three *Specularius* species have been observed on cultivated Phaseoleae.

1. Specularius albus (Pic). Described as from Sudan, *Specularius albus* has been found in Eritrea and beyond the Red Sea in Saudi Arabia.

In Sudan, it has been obtained from seeds of *Rhynchosia memnonia, Rhynchosia* sp. and amioga (?). In Eritrea, has been obtained from *Vigna* sp.

2. Specularius erythraeus (Pic). Rather widely distributed in Africa: Ghana, Nigeria, Cameroun, Tchad, Ethiopia, Eritrea, Kenya, Uganda, Tanzania, Burundi, Zaïre, Congo-Brazzaville, Angola, Zimbabwe, Malawi, Mozambique.

Has been obtained from *Cajanus cajan* seeds (= *indicus*) in Tanzania, Nigeria (Prevett, 1967) and in Kenya (Southgate, 1976) and from *Vigna unguiculata* in Zanzibar.

3. Specularius sulcaticollis (Pic.). Localized in the coastal area of East Africa: Somalia, Kenya, Tanzania, Mozambique.

Mentioned from *Vigna* sp. in Eritrea (Nastasi and Waldehaimonot, undated) and from *Cajanus cajan* in Kenya (Southgate, 1976).

E. Genus *Acanthoscelides* Schilsky

A genus essentially from the New World.

Acanthoscelides obtectus (Say). In Africa, the bean weevil is essentially distributed in the eastern mountainous areas: Uganda, Zaïr, Rwanda, Burundi, Kenya and Tanzania, in South Africa, Angola, Zimbabwe, in Madagascar and La Réunion. It is more rarely observed in hot areas (Cameroun, Congo-Brazzaville, Zaïre, etc. . .). The more ancient specimens that have been studied came from Madagascar (1898), Transvaal (1904), Tanzania (1911), Angola (1922). As early as the 1930s, it had well settled in the eastern mountainous areas. Its usual host-plant is of course the bean plant (*Phaseolus vulgaris*). Its attacks start in the fields on almost dry pods and as it is a polyvoltine species, they are carried on in stored seeds. In Swaziland, Walker (1979) mentions it in *Voandzeia subterranea* seeds. In Rwanda, it has been obtained from *Glycine max* (Soy). In Kivu, in the East Zaïre, a number of specimens hatched in 1949 from *Sesbania aegyptiaca* seeds.

2. SUBFAMILY *PACHYMERINAE*

Genus *Caryedon* Schönherr

A genus with a paleotropical distribution going into the eastern Mediterranean area as far as the Caspian Sea.

Caryedon serratus (Ol.). Distributed nearly all over tropical Africa, essentially in areas of arborean savanna (Mauritania, Senegal, Mali, Guinea, Guinea-Bissau, Ivory Coast, Ghana, High-Volta, Nigeria, Niger, Tchad, Cameroun, Sudan, Ethiopia, Kenya, Zaïre, Tanzania, Zimbabwe.

Its usual host plants are Caesalpinaceae: *Tamarindus indica, Piliostigma reticulata* and *thonningii, Bauhinia rufescens, Cassia sieberiana* and *arereh.*

This bruchid has adapted to the peanut by allotrophism (*Arachys hypogea,* of the Fabaceae family) in west Africa. Its attacks can start as early as the digging out and drying of pods and they are carried on in the peanuts kept in husks.

This species of African origin was introduced and has now well colonized many tropical areas consequently to the dissemination of one of its usual host-plants: *Tamarindus indica,* which in spite of its name originates from Africa.

3. SUBFAMILY *AMBLYCERINAE*

Genus *Zabrotes* G. Horn

A genus distributed in the New World.

Zabrotes subfasciatus (Boh.). This species of neotropical origin has been collected for quite a long time in African countries (Guinea-Bissau: 1899; Eritrea: 1902; Mozambique: 1914; Lower Zaïre: 1915; Tanzania: 1916). It seems to have well settled and spread considerably only in the past 20–25 years. It is now considered as being detrimental in Angola, Lower Zaïre, in East Africa (Kenya, Uganda, Tanzania, Burundi, Ethiopia). It is also frequently found in Madagascar and the Mascareigne Islands.

Its main host plant is the bean plant: *Phaseolus vulgaris.* It has also been observed on *Vigna* sp. in Uganda.

Out of these fifteen species detrimental to cultivated grain legumes observed in the Afrotropical area, 2 are of palearctic origin, 2 are of American origin, 2 are unquestionably of Asian origin, and 2 are of unprecise paleotropical origin. The other seven species are African ones.

The more harmful species at the present time are: *Bruchidius atrolineatus* at harvest time on *Vigna unguiculata: Callosobruchus maculatus* in the stored seeds of *Vigna*; *Acanthoscelides obtectus* on bean seeds, especially in mountainous areas of East Africa where this legume is a main contributor to the populations' food: *Caryedon serratus* on peanuts in husks in west Africa. Lastly, *Zabrotes subfasciatus,* detrimental to beans, has been spreading seriously these years, especially in East Africa. In addition to *Callosobruchus maculatus,* other species of the same genus can cause damage on various Phaseoleae.

Author's address:
Entomological Department of the Royal Museum of Central Africa
B-1980 Tervuren
Belgium

DISTRIBUTION, ECOLOGY AND IMPORTANCE OF BRUCHIDS ATTACKING GRAIN LEGUMES AND PULSES IN AFRICA

T. AJIBOLA TAYLOR
(Ibadan, Nigeria)

INTRODUCTION

The Bruchidae, made up of nearly 1400 species of bean and pea beetles, are major pests of seeds of both cultivated and wild leguminous plants throughout the tropics. Many of the species are well-known for the characteristic injury which their larvae cause in seeds of beans, peas and lentils, while others attack palm kernels, groundnuts and other oil seeds and nuts. Although they are frequently, and erroneously, referred to as 'weevils', they do not belong to the Curculionionoidea and do not possess the characters of weevils; they are better referred to as legume or bean beetles.

Bruchid infestation, in most parts of tropical Africa, starts in the field and, in most cases, continues in storage. This makes the family both store pests and field-to-store pests. In a few cases, field infestation soon terminates under storage conditions, e.g. in *Bruchidius atrolineatus* (Pic.) (Prevett, 1961), but in the majority of cases continuous infestation during storage leads to severe injury and considerable storage losses, as well as the appearance and establishment of exotic species through commerce and trade. For example, Southgate (1978) records three (*Callosobruchus analis* (F.), *Acanthoscelides obtectus* (Say) and *Zabrotes subfasciatus* (Boheman)) of the nine species of Bruchidae as introduced species in Africa; the other six species are indigenous and are now known to be widely distributed in Africa. Introduction and subsequent establishment obviously depend on optimum environmental conditions and availability of suitable hosts.

Field infestation by bruchids in Africa is characterised by its insidious nature. Eggs are usually laid on the drying pods with the young first instar larvae boring into the seeds. The seeds at threshing largely appear wholesome and undamaged, or damage may be as low at 2.5% (Caswell, 1968), but invariably they contain the developing larvae which serve as the major source of store infestation, and very often an apparently undamaged harvest becomes heavily infested in store after 4–8 weeks of storage, as is the case in *Callosobruchus maculatus* F. infestation. Rapid deterioration in storage in terms of weight and quality losses (Caswell, 1968; Taylor, 1969) have been recorded in Nigeria. Caswell (1968), estimated an annual loss of 24,000 tons of cowpea in Nigeria; this level of loss at present day production figures will be of the order of 500,000–600,000 tons of cowpea per year. Serious deterioration was noted by Parkin & Bills (1955) in the nutritive value of bruchid-infested legumes and from 48–70% damage has also been reported after 8 months by Lefevre (1950).

The depredations of bruchids on beans and pulses as major vegetable sources of protein assume considerable significance in Africa when viewed against the background of acute shortage of protein in most African diets and the resultant widespread protein malnutrition.

DISTRIBUTION OF BRUCHIDS IN AFRICA

The Bruchidae are represented in Africa by at least seven genera (*Callosobruchus, Bruchidius, Acanthoscelides, Caryedon, Pachymerus, Zabrotes* and *Spermophagus*). Of these genera, the genus *Callosobruchus* is most widely distributed and most destructive to legumes. *C. maculatus* (F.) and *C. chinensis* (L.) were reported to have become very widely spread, covering the whole of the African continent (Southgate, 1965) but more recently both *C. rhodesianus* (Pic.), previously restricted to the southern part of Africa, and *C. phaseoli* Gyll. have become more widespread, whilst *C. subinnotatus* (Pic.), a major pest of *Voandzeia subterranea* (L.) Thouars), has remained largely within its original area of distribution (Southgate, 1978).

Distribution of bruchids in Africa is influenced mostly by trade in beans, peas and pulses in all regions. In addition, accidental introduction occur through imported legume seed legume products from other parts of the world. The subsequent distribution of such species is influenced principally by optimum conditions for establishment and reproduction and dispersal through internal and inter-regional trade. *C. maculatus, C. phaseoli* (Gyll.) and *C. subinnotatus* are known to be indigenous and widespread in Africa. *C. rhodesianus,* another indigenous species, was initially restricted to its original home in the southern part of Africa until the 1950s but has now spread north of the equator and has been reported in Nigeria and other West and Central African countries (Booker, 1967). *C. chinensis* which is also regarded as indigenous, probably originated in Asia but had become established as a cosmopolitan species for many decades.

Both *Acanthoscelides obtectus* and *Zabrotes subfasciatus* have spread from their origins in South and Central America to Africa on imported stored products. Both species have become successful and widely distributed in the tropical and sub-tropical countries of West, North and Central Africa.

The four other genera of the Bruchidae are generally of less importance to food legumes. *Bruchidius atrolineatus* is a transient field pest of cowpea and other *Vigna* species, whilst other *Bruchidius* species have host plants such as *Acacia*, *Crotolaria* and *Sesbania* (Prevett, 1966). *Spermophagus* attacks *Ipomoea* and two species of *Pachymerus* are important as field pests of palm kernels in poorly-managed plantations of oil palm. *Caryedon serratus* is important because it attacks groundnuts (*Arachis*) an important food and cash crop of West and Central Africa. Other species of *Caryedon* also attack *Acacia* and *Cassia* species (Southgate, 1978).

Nearly all the species of food legumes grown in Africa, except some species of *Phaseolus*, are attacked by *C. maculatus;* it is followed in importance by *C. chinensis.* The cowpea appears to be the most suitable host for nearly all bruchids, and only *C. subinnotatus, A. obtectus* and *C. serratus* have so far not been recorded on it. *C. subinnotatus* can, however, infest *Arachis hypogea* (Prevett, 1970). Cowpea is the most important in food and nutrition in Africa; it is followed by pigeon pea, phaseolus beans, dolichos beans, soybeans. Groundnuts, soybeans and Bambara groundnuts (*Voandzeia*) are also important as food and oil seeds. Other genera and species of Bruchidae have been recorded on legumes in tropical Africa, including *Bruchidius baudoni* (Caillot), *Caryedon gonagra* (Fab.), *Caryedon albonotatus* (Pic.), *Caryedon cassiae* (Gyll.), *Pachymerus lacerdae* (Chev.) and other *Pachymerus* species, but the majority of these do not principally infest food legumes but wild host plants. They are generally considered of less importance.

C. MACULATUS – ECOLOGY AND IMPORTANCE

C. maculatus is obviously the most important in terms of distribution and losses in Africa. The biology of this most successful species has been studied by Larson and Fisher (1938), El Sawaf (1956), Howe and Currie (1964) and Booker (1967). Utida (1954) was the first to observe polymorphism and to describe a 'flight' or distributive form. Caswell (1956, 1960) reported the same form as an 'active' form from Nigeria. Later studies (Caswell, 1960; Utida, 1970, 1972; and Taylor, 1974a) showed that significant physiological, behavioural and reproductive differences exist between the 'active' and normal forms. Taylor (1975) discussed the significance of the active form as a mechanism for facilitating dispersal and field infestation. Recent studies (Taylor, in press) have shown that a readily flying 'active' form also exists in *C. subinnotatus,* a Bruchid which infests exposed seeds of *Voandzeia subterranea* from the field; and there are indications, from the examination of some characteristic differences in the pygidium of other field-to-store pest species of the Bruchidae, that this type of polymorphism is probably more widespread in the family. Temperature, humidity, initial population density, larval density, moisture content of seeds, continuous darkness and continuous light have been associated with the production of 'active' forms in *C. maculatus* (Sano, 1967; Utida, 1956, 1965, 1969, 1972; Taylor, 1974). Sano's (1967) suggestion that the active form was better adapted to life in the field fits into the role of 'active' females in field infestation of these field-to-store Bruchid pests. It can be further concluded that the so-called 'active' form is probably more primitive being the form originally associated with field-grown legumes; the normal, highly reproductive form, being a form which developed under conditions of abundance of hosts under storage conditions. Bruchid species which lack 'active' forms may be those which evolved long ago and may have lost their 'active' forms because store population densities were never high enough because of environmental conditions or effective storage control measures.

STORAGE METHODS AND THE FUTURE OF BRUCHID INFESTATION AND CONTROL

Although losses caused by Bruchids in stored legume seeds still constitute a major constraint in the availability of legumes for consumption in tropical Africa, storage methods conducive to effective disinfestation have emerged only slowly. Traditional storage of unthreshed cowpea, although capable of reducing overall damage (Caswell, 1968), has had only a marginal effect on legume losses. The major answer in most tropical areas where field-infestation is rampant is to shell soon after harvest fumigate with aluminium phosphide tablets in polythene protected sacks. Field infestation has remained high because of the proximity of storage sites to production sites. Commercialisation of processing, and fumigation combined with store hygiene should guarantee the availability of uninfested high quality legumes and pulses to the populations of Africans who for many decades will depend on these as major sources of vegetable protein.

REFERENCES

Booker, R.H., 1967. Observations on three burchids associated with cowpea in northern Nigeria. J. stored Prod. Res. 3: 1–15.

Caswell, G.H. 1956. Observations on the biology of *Callosobruchus maculatus* (F.) (Coleoptera: Bruchidae). Divl. Rep. Fac. Agric. Univ. Coll. Ibadan (Ent.) No. 3, 9 pp. multigraph.

Caswell, G.H., 1960. Observations on an abnormal form of *Callosobruchus maculatus* (F.). Bull. ent. Res. 50: 671–680.

Caswell, G.H. 1968. The storage of cowpea in the northern States of Nigeria. Proc. agric. Soc. Nigeria. 5: 4–6.

El Sawaf, S.K., 1956. Some factors affecting the longevity, oviposition and rate of development in the southern cowpea weevil, *Callosobruchus maculatus* (F.) (Coleoptera: Burchidae). Bull. Soc. ent. Egypt. 40: 29–95.

Howe, R.W. & Currie, J.C., 1964. Some laboratory observations on rates of development, mortality and oviposition of several species of Bruchidae in stored pulses. Bull. ent. Res. 55: 437–477.

Larson, A.O. & Fisher, C.K., 1938. The bean weevil and the southern cowpea weevil in California. Tech. Bull. U.S. Dept. Agric. No. 593, 70 pp.

Lefevre, P.C., 1950. Publ. Inst. nat. Etude Agron. Congo Belge Ser. Sci. No. 48, 65.

Prevett, P.F., 1961. Field infestation of cowpea (*Vigna unguiculata*) pods by beetles of the families Bruchidae and Curculionidae in northern Nigeria. Bull. ent. Res. 52: 635–645.

Prevett, P.F., 1966. Observations on the biology of six species of Bruchidae (Coleoptera) in northern Nigeria. Ent. mo. Mag. 102: 174–180.

Prevett, P.F., 1970. *Callosobruchus subinnotatus* (Pic.) (Coleoptera, Bruchidae) – a potential pest of stored groundnuts. J. stored Prod. Res. 6: 279–280.

Sano, I., 1967. Density effect and environmental temperature as factors producing the 'active' form of *Callosobruchus maculatus* (F.) (Coleoptera, Bruchidae). J. stored Prod. Res. 2: 187–195.

Southgate, B.J., 1965. Pulse bruchids in Africa. Proc. XII Int. Congr. Ent.: 642.

Southgate, B.J., 1978. The importance of the Bruchidae as pests of grain legumes, their distribution and control. In: Pests of Grain Legumes – Ecology and Control. van Emden, H; Singh, S.R. and Taylor, T. Ajibola (ed.). pp. 219–229.

Taylor, T. Ajibola, 1969. The effect of insect damage on the quantity and quality of Nigerian stored food crops. Contribution to the Nutrition Society of Nigeria Conference 1969.

Taylor, T. Ajibola, 1974. Observations on the effects of initial population densities in cultures, and humidity on the production of active females of *Callosobruchus maculatus* (F.) (Coleoptera, Burchidae). J. stored Prod. Res. 10: 113–122.

Taylor, T. Ajibola, 1975. The dynamics of active female production in *Callosobruchus maculatus* (F.) (Coleoptera, Burchidae) in stored cowpea. Proc. Inter. Working Cong. Stored Prod. Ent. 1: 209–218.

Taylor, T. Ajibola, in press. Studies on the biology of *Callosobruchus subinnotatus* (Pic.), a store pest of *Voandzeia subterranea* (L.) (Thouars), in Nigeria.

Utida, S., 1954. Phase dimorphism observed in the laboratory population of the cowpea weevil, *Callosobruchus quadrimaculatus* (In Japanese with English summary). Oyo – Dobuts Zasshi 18: 123–125.

Utida, S., 1956. Phase dimorphism observed in the laboratory population of the cowpea weevil, *Callosobruchus quadrimaculatus.* II: Differential effects of temperature, humidity and population density upon some ecological characters of the two phases. Res. Popul. Ecol. Kyoto Univ. 3: 93–104.

Utida, S., 1965. Phase dimorphism observed in the laboratory population of the cowpea weevil, *Callosobruchus quadrimaculatus.* IV: The mechanism of induction of the flight form. Jap. J. Ecol. 15: 193–199.

Utida, S. 1969. Photoperiod as a factor inducing the flight form in the population of southern cowpea weevil, *Callosobruchus maculatus.* Jap. J. appl. Ent. Zool. 13: 129–134.

Utida, S., 1970. Secular change of percent emergence of the flight form in the population of southern cowpea weevil, *Callosobruchus maculatus.* Jap. J. appl. Ent. Zool. 14: 71–78.

Utida, S. 1972. Density-dependent polymorphism in the adult of *Callosobruchus maculatus* (Coleoptera, Burchidae). J. stored Prod. Res. 8: 111–126.

Author's address:
Institute of Agricultural Research and Training
University of Ife
P.M.B. 5029
Ibadan
Nigeria

OBSERVATIONS ON THE ECOLOGY OF *BRUCHIDIUS ATROLINEATUS* PIC. AND *CALLOSOBRUCHUS MACULATUS* F (COLEOPTERA, BRUCHIDAE) IN NIGER

INEZDANE ALZOUMA
(*Niamey, Niger*)

INTRODUCTION

In the Republic of Niger, a country in the Sahel which has suffered from a severe food shortage over the last few years, cowpea – *Vigna unguiculata* L. Walp –, of which an estimated 300 000 tons are grown annually, might be important, granting its high nutritional value, in helping the local populations struggle towards self sufficiency in food resources (Smartt, 1976). However, this objective will stay beyond reach unless the very severe losses caused in the stores by two species of Bruchidae, *Bruchidius atrolineatus* Pic and *Callosobruchus maculatus* (F) are reduced. The larvae develop within the beans, feeding on the reserves in the cotyledons so that they are inedible. Those two species are associated, all over the southern part of the Republic of Niger. According to Caswell (1969) and Booker (1967) *Bruchidius atrolineatus* predominates in the fields and cannot breed in stores.

In Nigeria, Booker (1967) reports the presence of a third species, *Callosobruchus rhodesianus* Pic, which has not been observed in Niger. On the other hand, we have come across a few places infested with *Spermophagus humulus*, but we have failed so far to find the latter species in the stores.

Bruchid infestation of the cowpea fields usually begins late in August at the start of plant fructification when *Bruchidius atrolineatus* Pic and *Callosobruchus maculatus* F are seen about in considerable numbers. At this stage, both species are present on all plant organs while also hiding between the stamens of the spikes of Millet (*Pennissetum typhoides* Burm), or sorghum, (*Sorghum vulgare*). These plants are grown with cowpea, as often happens in Niger. These spikes may provide adequate protection against exposure to the sun and the adults may feed on their pollen during this period.

It is the purpose of this paper to survey the oviposition pattern of these two Bruchidae in a natural environment on the laboratory with respect to *Vigna unguiculata* pods, and to review our initial findings on the life cycle of these insects in the Niamey area in the Republic of Niger.

1. DISTRIBUTION OF *BRUCHIDIUS ATROLINEATUS* PIC EGGS LAID ON PODS OR BEANS OF *VIGNA UNGUICULATA* L. WALP

A. *A survey of pods collected in the field*

Methods. The study was carried out either in fields in the Niamey area (14 °NL) during the September harvesting time, that is, late in the rainy season, or on irrigated test plots cultivated at various periods in the year. When local variety was cultivated on the irrigated experimental plot flowing occurred 40 to 50 days after sowing. The first rip pods appeared 20 to 30 days thereafter. The pods harvested both in the field and in the experimental plot were observed regularly and the number of Bruchids eggs were noted. Ripe pods were harvested immediately; investigation was restricted to those pods from which only *Bruchidius atrolineatus* emerged.

Results. Females present on cowpeas or other plants can lay eggs on young green pods, as well as on dry ones. The eggs are glued on the pods. A survey of the distribution of these eggs on 80 dry pods shows that 80% of them are deposited along the sutures (Table 1). The egg axis always is parallel to the suture (Fig. 1A). Eggs are particularly abundant on that suture – usually the ventral one – less exposed to the sun.

Table 1. Study of the egg laying of *Bruchidius atrolineatus* on pods of cowpea collected in nature.

	Numbers	Percentages
Eggs on pods	1609 m = 20; 1 ± 2.7	–
Eggs along pods sutures	1313	84.1%
Eggs on pericarp of the pods	296	18.4%

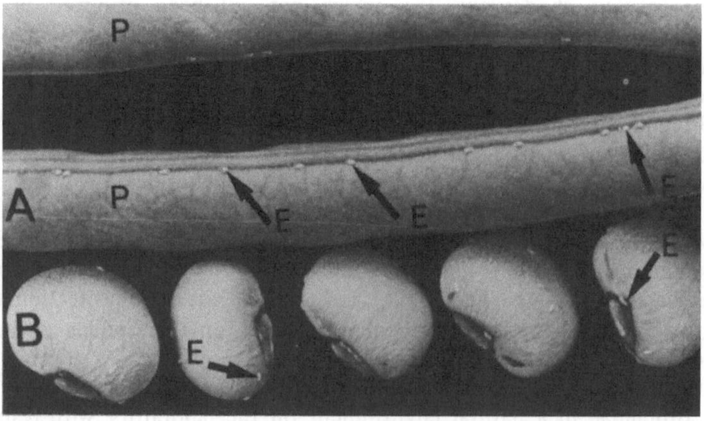

Fig. 1. Egg laying of *Bruchidius atrolineatus* on pod (A) or seed (B) of cowpea; S(suture); E(eggs); P(pericarp).

When *Bruchidius atrolineatus* lay their eggs on very young pods, the larvae develop in growing seeds and evolve into adults as the pods reach maturity. The first generation can then emerge from cowpea pods as soon as these are ripe. This accounts for the adults' emergence holes observed on seeds drying in the fields prior to the harvest.

B. *Study on pods contaminated in the laboratory*

Bruchidius atrolineatus, sexually paired soon after emergence, are put in test tubes sealed with fine-mesh gauze. Each tube contains a dry indehiscent pod of *Vigna unguiculata.* The tubes are kept in a rearing room whose climatic conditions are similar to those prevailing outdoors, i.e. II h L/13 h D; average temperature 27 °C in January 1980.

Under these conditions, life span ranges from 15 to 20 days. Oviposition starts 48 h after the experiment begins and reproduction extends over some ten days when no feeding takes place.

Table 2. Study of egg laying of *Bruchidius atrolineatus* Pic on 82 pods in laboratory.

	Numbers	Percentages
Pods	82	–
Eggs deposited	3619	–
on pods	m = 43.8 ± 40.4 on pods	
Number of eggs along the sutures	3449	95.3%
Number of eggs deposited on the pericarp of the pod	170	4.7%

As in the preceding case, the results reviewed (Table 2) show that a majority, 95% of the eggs are deposited along the pod sutures. However, the move from natural to laboratory conditions, in which a single pod and very little space are available to the female weevil, is followed by a markedly increase homogenization in spatial repartition. However, in both cases the number of eggs laids follows a normal distrubition (Fig. 2).

C. *Distribution of eggs laid on seeds*

When sexual pairs of *Bruchidius atrolineatus* are put on beans of *Vigna unguiculata* in the laboratory, the females are able to lay eggs. But these (Fig. 1B) are mostly stuck along the cotyledon sutures.

2. DISTRIBUTION OF EGGS LAID BY FLIGHTLESS *CALLOSOBRUCHUS MACULATUS* ON PODS OR BEANS OF *VIGNA UNGUICULATA*

Sexual pairs of flightless *Callosobruchus maculatus* (described by Utida, 1954) are put, soon after emergence, in test tubes sealed with fine-mesh gauze

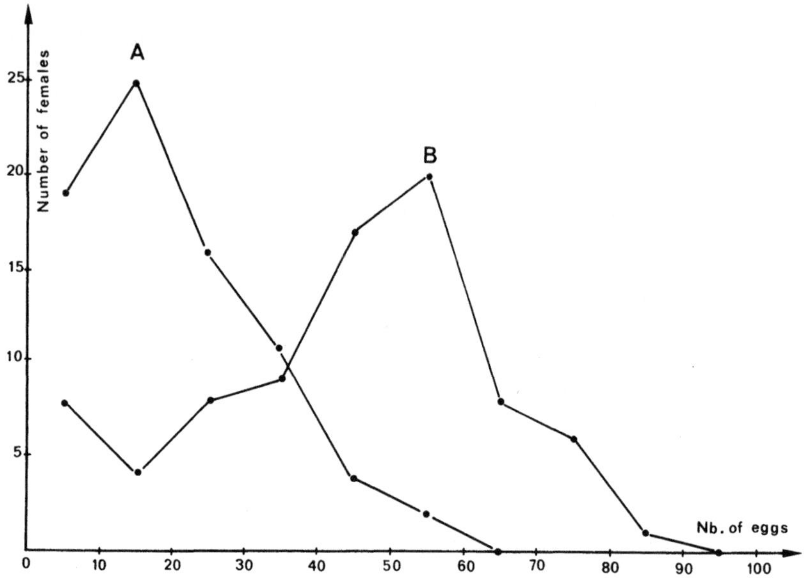

Fig. 2. Distribution of *Bruchidius atrolineatus* eggs deposited on pods in relation with the number of females. A – Eggs observed upon seeds collected in field. B – Eggs laid by females during experiment in laboratory.

containing dry ripe pods of *Vigna unguiculata.* The pods are observed at the end of the females' oviposition period, i.e., 15 to 20 days later. It is found then that oviposition starts in the first few days of the females' imaginal life, but, in this case, the eggs are glued on the lateral sides, either level to where a seed lies, or on the pericarp divide between two seeds (Fig. 3A). Only 2.5% of the eggs are deposited along the sutures (Table 3).

When pairs of these weevils are reared on seeds, the females lay their eggs all over, the tegument (Fig. 3B). Egg distribution then is totally different from that observed in *Bruchidius atrolineatus.*

Table 3. Study of egg laying of *Callosobruchus maculatus* on 50 pods in laboratory.

	Numbers	Percentages
Pods	50	–
Fecundity	1716 ($\bar{m}$ = 34.3 per female)	–
Eggs laid on the pods	1383 ($\bar{m}$ = 27 ± 4.4 per pod)	–
Eggs laid upon the pericarp	1354	97.4%
Eggs laid along the sutures	35	2.5%
Eggs laid upon the wall of the tube	327	–

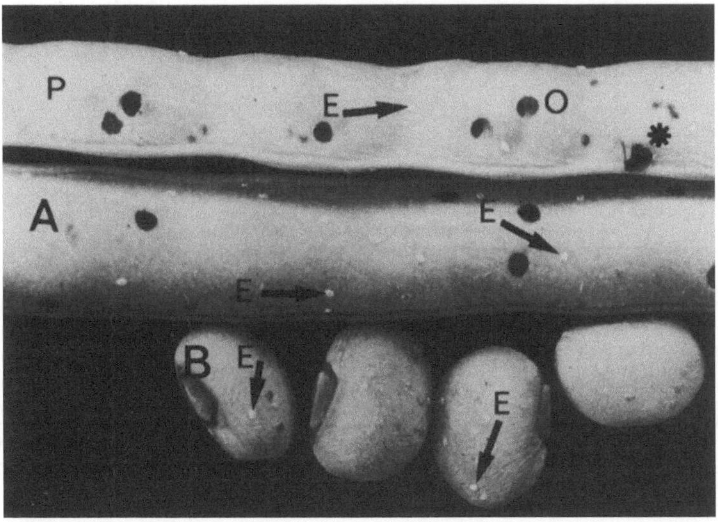

Fig. 3. Egg laying of *Callosobruchus maculatus* on pod (A) and seed (B) of cowpea. (P) pericarp (E) Egg (O) orifice of emergence of adults (∗).

3. EGG DISTRIBUTION WHEN FEMALE *CALLOSOBRUCHUS MACULATUS* ARE IN PRESENCE OF A DEHISCENT POD AND A SEED OF *VIGNA UNGUICULATA*

Sexual pairs of flightless *Callosobruchus maculatus* are put (Fig. 4) in rearing boxes containing three beans and one pod of *Vigna unguiculata.* The experiment proceeds under the same thermo and photo-periodical conditions as previously. Statistical analysis of the results obtained with 72 pairs shows that 84% of the eggs are deposited on seeds (Table 4).

All the seeds put inside the boxes had eggs deposited on them. Only 59.7% of the pods have received eggs – very few in many cases.

Thus, flightless females of *Callosobruchus maculatus* will rather deposit eggs on beans when they are in presence of a pod and seeds of *Vigna unguiculata.*

4. PRELIMINARY OBSERVATIONS ON THE BIOLOGY OF *CALLOSOBRUCHUS MACULATUS* F. AND *BRUCHIDIUS ATROLINEATUS* PIC IN THE DRY SEASON IN NIGER

(1) During the dry season, *Bruchidius atrolinatus* Pic and *Callosobruchus maculatus* F. can be developed in storages – cowpeas are stored as pods or seeds in traditional granaries or in jute sacks by merchants (Fig. 5) and storage organisations. Insect in storage could contaminate crops in the field as has been observed in Europe for *Acanthoscelides obtectus* (Arnoux & Labeyrie, 1957; Jarry, 1981) but there is a possibility that some insects

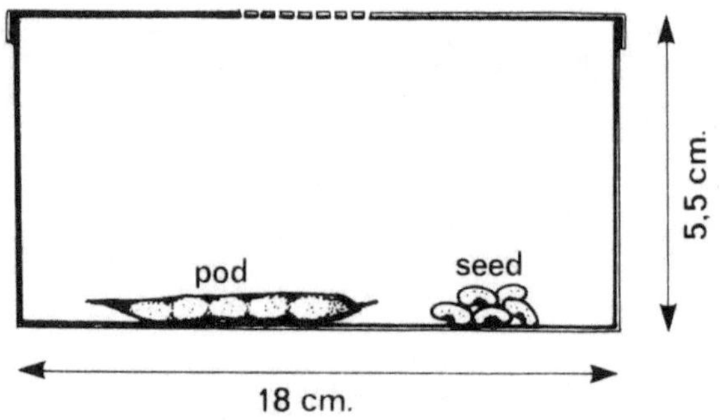

Fig. 4. Rearing box where choise experiment where made.

Table 4. Repartition of eggs when females are in presence of seeds and pods of cowpea.

	Numbers	Percentages
Pairs of *Callosobruchus maculatus* F.	72	–
Pods	72	–
Seeds	216	–
Eggs deposited on seeds	2366	83.9%
Eggs deposited on pods	289	10.2%
Eggs deposited on the wall of the rearing box	163	5.7%
Total number of eggs	2818	100%

Fig. 5. Traditional granaries in the field in Niamey country (Niger).

can outlive the dry season in the field as is suggested by crops grown in this season on irrigated test plots remote from storage areas.

Thus, cowpea grown as from October 1979 has provided us with evidence of investation by *Bruchidius atrolineatus* Pic in the dry season.

A very high proportion of the plants that fructified in December were contaminated – 100% of the harvested pods have an average on 20 eggs per pods. Contamination decreased in February and March (Table 5) and the number of eggs deposited on each also decreased, range from 1 to 5. Thus it seems that the more adverse the weather conditions get, the fewer and the less fecund *Bruchidius atrolineatus* Pic become.

Table 5. Contamination of cowpea pods by *Bruchidius atrolineatus* during the dry season.

	Rate of contamination of green pods	Rate of contamination of ripe pods
February 1980	2.9%	47.3%
March 1980	2.3%	6.5%

(2) In late December 1979, 24 pods of a wild (*Vigna unguiculata Suf. dehindtiana* (harms)* were collected in a irrigated zone in the Niamey region. These pods, infested with eggs whose distribution was that of *Bruchidius atrolineatus* Pic, were kept under observation in the laboratory. Between the 9th January and the 29th of March 1980, we recorded the emergence of 24 *Bruchidius atrolineatus* Pic (14 females, 10 males) against two flightless *Callosobruchus maculatus* F. only.

Sexual pairs formed from that population of *Bruchidius atrolineatus*, put in test tubes containing cowpea pods, failed to lay eggs; they seemed to present an imaginal diapause analogous to that described in wild populations of *Acanthoscelides obtectus* from Mexico (Biemont & Bonet, 1981).

Existence of populations of *Bruchidius atrolineatus* on another species of wild Vigna (*Vigna triloba*) has already been reported in Nigeria.

According to Booker (1967), Center & Johnson (1974) a variety of leguminous plants are suitable for the development of *Callosobruchus maculatus* F. But, out of 34 species of pulses whose pods we have examined in Niger, only wild and cultivated *Vigna unguiculata* have proved to be host plants to *Callosobruchus maculatus* F. and *Bruchidius atrolineatus* Pic.

(3) 25 cowpea pods left over from the harvest in October were collected in February 1980 at Tarna (Maradi country in the East of Republic of Niger.) In this zone a herbaceous and arborescent savanna climatics conditions are similar to those of Niamey but temperatures are lower.

These pods were brought to the laboratory and we recorded the emergence of 113 insects one month later. None of the sexual pairs of *Bruchidius atrolineatus* laid eggs on pods or seeds.

(4) A more detailed study was attempted in the laboratory concerning the reproduction of *Bruchidius atrolineatus* adults emerging from 80 contaminated dry pods collected early in January 1980.

284 adults emerged from the pods between 21st of January and the 1st of March 1980; their reproductive activity was recorded as a function of their time of emergence.

*We thank Doctor R. Marechal (Faculté des Sciences Agronomique de Gembloux) for the help in determination of Vigna species.

25 sexual pairs, formed from individuals emerging late in January were put in rearing boxes containing ripe pods. It was then found that 21 females (84%) had mated and laid eggs.

25 pairs were formed from individuals emerging late in February. These insects had had a longer post-embryonary development, or were evolved from eggs deposited on the pods more lately. It was then found that 21 females (84 (84%) do not lay any eggs although placed in conditions favourable to oviposition. This points to the possible existence of imaginal diapause in insects emerging late – that is at a time when weather conditions are particularly drastic – due to the dry season enabling them to survive until the plants bear fruit.

Diapause of this type – in the dry season – is frequent in insects living in the tropical zone (Scheltes, 1978); it can occur either at the adult stage (Norris, 1965) or at various larval or nymphal stages (Scheltes, 1978).

DISCUSSION

This preliminary research made on *Bruchidius atrolineatus* and *Callosobruchus maculatus* in Niger shows that these two insects have different oviposition patterns in a natural environment.

Bruchidius atrolineatus, which sticks its eggs along the pod sutures, can lay them on green pods as soon as these are formed. The first generation is able to develop through the larval stages in growing seeds and evolve into adults when the pods are ripe. The adults then deposit their eggs on the dry pods or on beans that have just been stored. The second and the next generations will then develop in the granary, which accounts for the extensive damage at this stage.

Callosobruchus maculatus in a natural environment lays eggs on dry pods, gluing them all over the pericarp. Cowpea being stored in the granaries as pods or beans, the emerging adults will be placed in conditions suitable for their development.

There is a strong case, then, for tracing back to where they belong the weevils infesting cowpea crops in the field. Some preservation techniques in Niger may facilitate that contamination – for instance, non-insect proof wattle and daub granaries erected in the fields, or the drying and storage of cowpea leaves and pods on trees. But the insects can probably survive the dry season in a natural environment until growing or ripening pods enable them to breed. Cowpea crops grown during dry season evidence damage in the field even during very hot spells. The insects may find in a natural environment host plants suitable for their development, as reported by Booker (1967) and the present author, or enter imaginal diapause. Investigation of this phenomenon is now in process. For it is by development on knowledge of these two Bruchidae in a natural environment that one will be able to work out some really efficient methods of controlling them.

REFERENCES

Arnous, J., Labeyrie, V. & Maison, P., 1957. Influence de la temperature et de l'isolation sur la migration des bruches du haricot (*Acanthoscelides obtectus*) dans la nature. CR. Acad. Sc. 247: 2343–2345.

Biemont, J.C. & Bonet, A., 1981. Preliminary observations of bean weevil populations from the *Acanthoscelides obtectus* Says. Group living on wild or subspontaneous *Phaseolus vulgaris* L. and *Phaseolus coccineus* L and on *Phaseolus vulgaris* cultivated in the Tepoztlan region state of Morelos – Mexico, this volume.

Booker, R.H., 1967. Observations on three Bruchids associated with cowpea in Northern Nigeria. J. Stored Prod. Res. 3: 1–15.

Caswell, G.H., 1961. The infestation of cowpeas in western Region of Nigeria. Trop. Sci. 3: 154–158.

Center, T.D. & Johnson, C.D., 1974. Coevolution of some seed beetles (Coleoptera Bruchidae) and their hosts. Ecology 55: 1096–1103.

Jarry, M., 1981. Evolution of spatial pattern of attacks by *Acanthoscelides obtectus* Say (Coleoptera: Bruchidae) *Phaseolus vulgaris* in south west France, this volume.

Norris, M.J., 1965. The influence of constant and changing photoperiods on imaginal diapause in the red locust (*Nomadacris septemfasciala*). J. Ins. Physiol. 2: 1105–1119.

Ouedraogo, P.A., 1978. Etude de quelques aspects de la biologie de *Callosobruchus maculatus* et de l'influence des facteurs externes stimulants (plante hôte et copulations) sur l'activité reproductrice des femelles. Thèse 3e cycle, Université de Toulouse, 101 pp.

Scheltes, P., 1978. Ecological and physiological aspects of aestivation diapause in the larvae of two Pyralid stalk borers of maize in Kenya. Thesis, University of Wageningen, 109 pp.

Smartt, J., 1976. Tropical pulses. Longman 348 pp.

Utida, S., 1954. Phases dimorphism observed in the laboratory population of the cowpea weevil *Callosobruchus maculatus*. Jap. J. Ecol. 18: 161–168.

Author's address:
Université de Niamey
Laboratoire de Biologie
B.P. 10662
Niamey
Niger

DISTRIBUTION AND IMPORTANCE OF BRUCHID ATTACKS ON DIFFERENT SPECIES OF PULSES CONSUMED IN THE NEAR EAST

G. HARIRI
(Aleppo, Syria)

INTRODUCTION

The term 'Near East' is referred to the lands around the eastern end of the Mediterranean Sea. The region is characterized by a typical Mediterranean-type climate which is winter rainfall (or snow in the high lands) alternating with summer drought when evapotranspiration exceeds precipitation for several months with considerable variability in climate in the region as a result to the influence of the Mediterranean, Black and Caspian Seas and the plateaus of the region. In addition to the amount and distribution of rainfall which determine the growing period of rainfed crops, temperature of winter and spring influence the production. The interaction of altitude, latitude and distance from the Seas evolve the agroclimatic sub-regions within the Mediterranean climate to semi-arid continental climate of the high plateaus of Central Turkey, Northern Iran and Afghanistan. Accordingly, the Near East can be tentatively divided into three sub-regions: High Plateau Areas (Iran, N. Iraq and Turkey), Lowland Areas (Syria, Lebanon and Jordan) and Southern Areas (Egypt, S. Iraq and S. Iran).

FOOD LEGUMES OF THE REGION

Food leguminous crops are important source for human nutrition and to agricultural systems in the Near East. They are essential component of rainfed agricultural system in the region; and commonly grown in rotation with cereals.

There are two types of rotation under rainfed conditions; each has its winter and summer crops. Wheat, barley and winter legumes (lentils, broad-beans and vetchs) are grown in winter. Sorghum, sesame, watermelon and spring legumes (chickpeas) are grown in spring. Beans, cowpeas, peas, peanut and soy beans are grown under irrigation as summer crops.

The predominant farming system in the higher rainfed area is a three-year rotation of cereals, legumes and fallow or summer crops. In Egypt, grain legumes are grown under irrigation as winter crops in 2-year rotation with cereals or cotton. Table 1 shows the pulses production of the region.

Series Entomologica, Vol. 19: ed. by V. Labeyrie

Table 1. Selected data on the Near East pulses production (1000 MT), annual average, 1977 (FAO, Statistical Book).

	Dry beans	Dry broad beans	Dry peas	Dry cowpeas (1975)	Chickpeas	Lentils	Soy beans	Groundnut in shell	Total
Cyprus	1	3						1	5
Egypt	18	237	6	9	6	45	15	35	371
Iran	111		30		40	30	103		314
Iraq	5	28		3	9	6	1		52
Jordan						7			7
Lebanon	1	1			3	2		4	11
Syria	12	12	1		55	117		24	221
Turkey	160	48	4		168	118	8	52	558
Total	308	329	41	12	281	325	127	116	1539

BRUCHIDS ATTACKING FOOD LEGUMES

Bruchids attacking food legumes in different countries of the Near East have been reviewed by several authors (e.g.: Alkan, 1966; Avidov & Harpaz, 1969; Bishara *et al.*, 1967; F.A.O., 1977; Gentry, 1965; Hariri, 1979, 1980; Kalkan, 1973; Ozer, 1961; Rivnay, 1962; Shomar, 1963). The most important species, their geographical distribution and legumes hosts are shown in Table 2.

Table 2. Major bruchid species attacking pulses in the Near East.

A. Species producing one generation a year		
Bruchus dentipes Baudi	Eastern Mediterranean to Afghanistan	Broadbean, vetch, other legumes.
B. emarginatus All.	S. Europe, Near East, N. Africa, Asia	Peas, cowpea, vetch, other legumes.
B. ervi Fröel.	Europe, N. Africa, Near East	Lentil, beans vetch, other legumes.
B. lentis Fröel.	Europe to Central Asia, Near East, India	Lentil, vetch.
B. pisorum L.	Cosmopolitan	Peas, broad-bean.
B. rufimanus Boh.	Europe, N. Africa, Near East	Broadbean, beans, peas, vetch, lentil.
B. signaticornis Gyll.	Europe, N. Africa, Turkey, Syria	Lentil.
B. tristis Boh.	Europe, Near East	Vetch, peas, beans.
Bruchidius quinqueguttatus (Oliv.)	Mediterranean	Broadbean, lentil, vetch, chickpea, others.
B. Species producing several generations a year		
Callosobruchus chinensis (L.)	Cosmopolitan	Cowpea, chick-pea, pea, lentil others.
C. maculatus (F.)	Tropical and Subtropical	Cowpea, chick-peas, others.
Acanthoscelides obtectus (Say)	Cosmopolitan	Beans, lentil chickpea, others.
Bruchidius incarnatus (Boh).	N. Africa, Egypt	Broadbean, peas, chick-pea, lentil.

IMPORTANCE OF BRUCHIDS IN THE REGION

A. Bruchus species

Informations about the importance of bruchids as pests of food legumes in the Near East show that lentils, broadbeans and peas are the legume crops attacked annually by *Bruchus* species. The degree of infestation differs from year to year and from one country to another.

The most common and serious *Bruchus* species attacking lentils are lentil seed beetle, *B. ervi*, European lentil seed beetle, *B. lentis* and the Black Sea lentil seed beetle, *B. signaticornis* causing infestation up to 80% of lentil seeds in Syria, Turkey and Iran.

The large broadbean seed beetle, *B. rufimanus* infests broadbean seeds causing serious damage in Lebanon, Syria, Iran and Egypt. The infestation may reach in Egypt up to about 20% (Bishara, 1967). The infestation level may reach up to 40–50% of the seed (Avidov & Harpaz, 1969). Recent estimation of broadbean seeds infestation by *B. dentips* at ICARDA field showed that local and small cultivars infestation may reach up to 70% (ICARDA, unpublished).

The pea seed beetle, *B. pisorum,* attacks peas in the region; it is abundant where peas are grown for the dry seeds. In Iraq, *B. pisorum* is considered as a pest of broadbean (al-Rawy & Kaddow, 1966).

Other *Bruchus* species infesting these major legume crops in the region are of minor importance.

Biological studies on the previous *Bruchus* species indicated that over-wintered adults emerge from hibernation in spring to feed on pollen; mate and lay their eggs on pods. Upon hatching, larva bores for feeding into the seed completing its development. Pupation takes place in the seed. The larval duration and the time of adult emergence from the seed differ according to different *Bruchus* species and to the legume seed hosts. In Egypt, *B. rufimanus* has been found the only bruchid species attacking broadbeans in the field. The adult beetle leaves the broadbean seed after being sown in wet medium. The eggs are laid on pods during February–March. Larval and pupal development to reach the adult stage takes about 7 months. The adult remains inside the broadbean seed until sowing time in October–November. In stored seeds, the adult remains inside the seed until next April after which it dies without leaving the seed (Bishara *et al.*, 1967). *B. pisorum* when infesting broadbean seeds in Iraq, the adults do not emerge from the dry infested seeds in the storage until the seeds are soaked in water. The source of field infestation is the previous season's seeds that are planted or from those left in the field after harvesting. The seeds that are kept in storage for two successive summers will insure 100% mortality (al-Rawy & Kaddow, 1966). On the other hand, *B. pisorum* infesting peas in the region develops its larval and pupal stages in the spring. Newly emerged adults leave the pea seeds in early summer for aestivohibernation in suitable shelters emerging from hibernation next early spring (Rivnay, 1962). *B. ervi* infesting lentil seeds, the adult leaves the seed in June or July and re-enters the lentil seed again for aestivohibernation, re-appearing in next spring when the average daily temperature exceeds 10 °C (Dortbundak, 1975).

B. Callosobruchus species

The most common and serious species of the genus *Callosobruchus* in the Near East are the cowpea seed beetle, *C. maculatus* and the Adzuki bean seed beetle, *C. chinensis.* These beetles are major storage pests of grain legumes attacking seeds of beans, peas, chickpeas, lentils, cowpeas and broadbeans. They raise serveral generations per year; accordinly considered more destructive. The larvae of both species feed on the seed contents. The heavy infestation resulting in unsuitable seeds either for planting or human consumption.

C. maculatus could reduce the weight of stored cowpea seeds to over half during three months of a heavy infestation (El Sawaf, 1956).

C. chinensis has been reported infesting about 70% of stored lentils in Cyprus (Georghiou, 1977). Testing results of 700 lentil entries at room conditions to infestation by *C. chinensis* showed that 32 to 94% of lentil seeds were infested with a weight loss ranging from 8 to 46%. Infested large and small lentil seeds by *C. chinensis* gave germination percentages for seed harbored one bruchid 84 and 25; two bruchids 32 and 5; three bruchids 20 and 0, respectively; for 4, 5 and 6 bruchids in each large lentil seed gave 11, 8 and 2, respectively, while infested small and large lentil seeds by *B. ervi* gave 0 and 20% germination, respectively (Hariri, 1980).

The optimum condition for rearing *C. maculatus* are 25–30°C (el-Kifl, 1975). Rearing of *C. maculatus* at 15 °C resulted that fewer eggs were laid and none hatched; breeding of 4 generations at 18 °C did not result acclimatisation to low temperatures (Conen & Calderon, 1977).

Many studies were carried out in the region on food preference, and legumes resistance to *C. chinensis* and *C. maculatus* infestation. *C. chinensis* completed its development on broadbean faster than *C. maculatus* (Ahmed *et al.*, 1978). In recent laboratory studies in India, lentil was found unsuitable food for *C. maculatus*. *C. chinensis* prefers cowpea followed by lentil, chickpea and pea; while *C. maculatus* prefers cowpea, followed by chickpea, pea and lentil (Singh, 1976; Gokhale, 1973; Yadav & Pant, 1978). Laboratory tests which carried out in Egypt on the oviposition preference of *C. maculatus* showed that cowpea was the most preferred for oviposition followed by beans, broadbeans, soy beans and lentils; the greatest loss of weight of seeds infested for five months was for cowpea (about half) and least for lentil (4.5%) (Koura *et al.*, 1971).

Studies on the resistance of chickpeas to attack by *Callosobruchus* have shown that varieties with rough-textured or spiny seed coats appear to act as deterrants to oviposition and thus are less heavily infested than smooth coated varieties (Raina, 1971; Schalk, 1973; Schalk *et al.*, 1973).

The evidence of the association between colour of seed coat of broadbean seeds and infestation by *C. maculatus*, higher infestation was obtained on brown seeds comparable to lower infestation on white seeds (Abo-Hegazi *et al.*, 1978).

The resistance could be due to the chemical consitution of the seed or seed coat colourant or the physical properties such as coat thickness (Abo-Hegazi *et al.*, 1978; Applebaum *et al.*, 1965; Podoler & Applebaum, 1968). It was found the wrinkled peas starch is higher in amylose which is of poor nutritional value to *C. chinensis* compared with smooth peas varieties (Podoler and Applebaum, 1971).

C. *Bruchidius species*

The small broadbean beetle, *B. incarnatus*, causes serious damage to broadbean seeds in store in Egypt. It may infests peas, cowpea and lentils. The insect raises several generations a year. At room temperature 7, 8 and 9 generation a year were obtained when the beetle was provided with broadbeans, peas and lentils, respectively (el-Kifl & Metwally, 1971). The duration from egg to adult may be completed in about 45 days at 26 °C and 70% R.H. *B. quinqueguttatus* has been recorded infesting lentil seeds from Northern Syria. This beetle produces one generation a year. (ICARDA, unpublished).

D. *Acanthoscelides obtectus*

The bean seed beetle, *A. obtectus*, feeds on most varieties of beans (*Phaseolus*) in storage, but also attacks broadbeans, lentils and chickpeas. The insect produces at least 6 generations a year.

REFERENCES

Abo-Hegazi, A.M.T., Nassib, A.M. & Ahmed, M.Y.Y., 1978. The association between seed coat colour and brightness of field bean (Vicia faba L.) and resistance to legume beetles. Proc. 4th Conf. Control, Cairo, 768–778.

Ahmed, S.E., Kamel, A.H. & Wahab, A.E.A., 1978. A study of the duration of the different stages of Callosobruchus chinensis L. and Callosobruchus maculatus F. under constant conditions (Coleoptera, Bruchidae). Agric. Res. Rev. 56: 151–154.

Alkan, B., 1966. Türkiyenin Zarali tohum böcekleri (Coleoptera-Bruchidae) fauna'si üzerinde Calişmalar. Ankara Univ. Ziraat Fak. Yayin, 277 Calis. 174, 56 pp.

Applebaum, S.W., Gestetner, B. & Birk, Y., 1965. Physiological aspects of host specificity in the Bruchidac IV. Developmental incompatability of Soybeans for Callosobruchus. J. Insect Physiol. 11: 611–616.

Avidov, Z. & Harpaz, I., 1969. Plant pests. 549 pp. (Bruchidae, 281–290).

Bishara, S.; Haggag, Y.M. & Riad, A.A., 1967. Field infestation of broad beans by bruchids in U.A.R. Agric. Res. Rev. 45: 33–39.

Conen, M. & Calderon, M., 1977. Effects of continuous rearing at low temperature on the reproduction of Callosobruchus maculatus (F). Prog. Rep. Stored Prod. Div. pp. 5–6 (R.A.E. 66, abs. no. 399).

Dortbudak, N., 1975. Mardin ili mercimeklerine ariz olan Bruchus ervi Fröhl. un biyoekolojisi ve mücadele metodlari uzerinde arastirmalar. Turk. Cum. Gida-Tarim Hayvan Bakanligi, Aras. Eser. Serisi NO 39, 46 pp.

F.A.O., 1977. Important Storage pests of food legumes. In: Food legume crops: improvement and production. FAO P1. Prod. prot. paper 9: 93–97.

Gentry, J.W., 1965. Crop insects of Northeast Africa-Southwest Asia. 210 pp. (Bruchidae, pp. 178–180). USDA, Agric. Handbook No. 273.

Georghiou, G.P., 1977. The insects and mites of Cyprus. 347 pp. (Bruchidae, pp. 36–37). Kiphissia, Athens.

Gokhale, V.G., 1973. Development compatibility of several pulses in the Bruchidae-I. Growth and development of Callosobruchus maculatus (Fabricius) on host seeds. Bull. Grain Tech. 11: 28–31.

Hariri, G., 1979. Insect pests of chickpea and lentil in the countries of the Eastern Mediterranean: A review. pp. 120–123. In: Food legume improvement and development. Edited by G.C. Hawtin & G.J. Chancellor. 216 pp. IDRC & ICARDA.

Hariri, G., 1980. Insect pests of lentil. In: Lentil Book (In the press). C.A.B. (London).

Kalkan, M., 1973. Investigation on the species, distribution and percentage of damage of Bruchidae in Central Anatolia. Ankara Bolge Zirai Muc. aras. No. 64: 192.
el-Kifl, A.H. & Metwally, M.M., 1971. Biological and ecological studies on Bruchidius incarnatus Boh. (Coleoptera: Bruchidae). Bull. Soc. ent. Egypte 55: 141–162.
Koura, A., el-Halfawy, M., Shehata, T. & Darkal, W., 1971. Preference of the cowpea weevil, Callosobruchus maculatus F., to some legume seeds and weight loss due to insect infestation. Agric. Res. Rev. 49: 35–40.
Ozer, M., 1961. Researches on some pests of stored grain and products in Turkey. Univ. Ankara Yearbook, Fac. Agric. pp. 81–95.
Podoler, H. & Applebaum, S.W., 1968. Physiological aspects of host specifity in the Bruchidae. V. Varietal differences in the resistance of Vicia faba L. to Callosobruchus chinensis L. J. Stored Prod. Res. 4: 9–11.
Podoler, H. & Applebaum, S.W., 1971. Host specificity in the Bruchidae. VII. The effect of carbohydrate composition on varietal resistance of garden peas to Callosobruchus chinensis L. J. Stored Prod. Res. 7: 97–105.
Raina, A.K., 1971. Comparative resistance to the species of Callosobruchus in a strain of chickpea (Cicer arietinum L.) J. Stored Prod. Res. 7: 213–216.
al-Rawy, M.A., & Kaddow, I.K., 1966. Preliminary observations on the control of the Pea weevil Bruchus pisorum L. attacking the broad beans Vicia faba L. Bull. Biol. Res. Cent. 2: 86–87.
Rivnay, E., 1962. Field crop pests in the Near East. 450 pp. (Bruchidae pp. 212–216. Junk, The Hague.
el-Sawaf, S.K., 1956. Some factors affecting the longevity, oviposition and rate of development in the cowpea weevil Callosobruchus maculatus F. Bul!. Soc. ent. Egypte 40: 29–95.
Schalk, J.M., 1973. Chickpea resistance to Callosobruchus maculatus in Iran. J. econ. Ent. 66: 578–579.
Schalk, J.M., Evans, K.H. & Kaiser, W.J., 1973. Resistance in lines of chickpeas to attack by Callosobruchus maculatus in Iran. P1. Prot. Bull. FAO 21: 126–131.
Shomar, N.F.H., 1963. A monographic revision of the Bruchidae of Egypt (U.A.R.) (Coleoptera). Bull. Soc. ent. Egypte 47: 141–196.
Singh, Y., 1976. Studies on relative resistance of important pulses to Callosobruchus maculatus (Fabricius) and Callosobruchus chinensis (Linnaeus). Ent. Newsletter 6: 18–19.
Yadav, T.D. & Pant, N.C., 1979. Developmental response of Callosobruchus maculatus (Fab.) and C. chinensis (Linn.) on different pulses. Ind. J. Ent. 40: 7–15.

Author's address:
Faculty of Agriculture
University of Aleppo
Aleppo
Syria

[illegible]
[illegible] Agric. Res. [illegible]
[illegible]

THE PARASITES OF BRUCHIDS

J.R. STEFFAN
(Paris, France)

It is not an easy work to deal with parasites of *Bruchidae* for until now, apart from some exceptional cases, these insects have hardly aroused any interest among taxonomists or specialists in applied entomology. There is little information available in the literature and what exists is often inaccurate. In general, all that can be learnt is that such and such a Hymenoptera grows at the expense of such and such a *Bruchidae.* Thus, in the United States, one can find 17 species of Eulophids belonging to the genus *Horismenus,* of which three are parasites of *Mimosestes, Stator* and *Acanthoscelides.* Yet although some *Horismenus* have brought about major reductions in the *A. obtectus* (Say) populations in Mexico (unpublished research carried out by I.B.E.A.S.) not a single paper has been written about any one of them. Meanwhile, in North America twelve papers have been published on the biology of *H. floridanus* (Ashm) and *H. fraternus* (Fitch) which attack orchard Microlepidoptera and their parasites, the *Apanteles.* The *Bruchidae* as well as their natural enemies are poorly considered.

The best way of getting some idea of the entomofauna related to the *Bruchidae* is to consult descriptive catalogues. The most important is that published by De Luca in 1965, an authoritative work covering the five continents which is presented in three parts. It contains a list of the various *Bruchidae* with their corresponding parasites and predators, a list of these parasites and their respective hosts and finally a country-by-country index. Also worth mentioning are Peck's excellent catalogue on neartic Chalcidoidea brought out in 1963 and the huge work on Hymenoptera in North America – except for Mexico – which was published by Krombein *et al.*, (1979). Of course, the information obtained from these works is not always consistent: *Acanthoscelides horni* (Pic) cited as the host of *Zatropis orontas* (Walk) (*Z. bruchivorus* Ashm) in the work of Krombein *et al.* is not mentioned by Peck. On the contrary, *Bruchus brachialis* Fähr., cited by Peck as being the host of *Eurytoma tylodermatis* Ashm is excluded by Burks in the Krombein catalogue. It also happens that a species left out in the text of the book is only included in the index of hosts such as *A. submuticus* (Sharp) victim of *Eupelmus bruchivorus* (Crawf.) (see Peck, p. 479 and p. 956). So, the reader should be cautious. Indeed, numerous question marks found in the lists given by these entomologists are a warning by themselves.

Series Entomologica, Vol. 19: ed. by V. Labeyrie

Of all the catalogues mentioned, De Luca's is the most extensive. For the time being, a commentary on De Luca's catalogue (while ignoring questionable parasites) seems to be the best way of meeting the aim of this paper.

The first thing to emerge from the work we owe to De Luca is the disproportion between the number of *Bethyloidea* and *Ichneumonoidea* – less than twenty species altogether – and the number of *Chalcidoidea* – about seventy species.

Among the five Bethylids mentioned, only *Sclerodermus immigrans* (Bridw.) and *Laelius utilis* Cock should be retained.

Among the *Ichneumonoidea,* no *Ichneumonidae* can be found. The only species mentioned in De Luca's lists – *Calliephialtes notandus* (Cress) – attacks only Lepidoptera.

The *Bruchidae* are parasitized by about fifteen species of *Braconidae,* most of them belonging to the three genera *Triaspis, Heterospilus* and *Urosigalphus* (s.g. *Bruchiurosigalphus*).

Within the superfamily *Chalcidoidea*, the family *Chalcididae* s.tr. includes only a few lariophagous species. They are usually bigger than other Chalcids and they require hosts larger than that of most larvae of *Bruchidae* to complete their growth. Thus it is not surprising if *Antrocephalus mytis* (Wlk.) (*A. aethiopicus* Ms) only attacks *Caryedon serratus* (01). *Euchalcidia caryobori* Hanna parasitizes *Caryoborus pallidus* (01) and also the previous Bruchid in Senegal (personal observation). It is not, however, certain whether *Hockeria unicolor* Wlk. has *Spermophagus sericeus* (Geof) as host. *Semiotellus mundus* Wlk. (*Semiotus varians* Wlk.) included in the *Chalcididae,* is in fact a Pteromalid the biology of which is still unknown.

Within the family *Encyrtidae*, the *Habrolepoidea* are without question parasites of the *Bruchidae.* According to Burks (in Krombein *et al.*, 1979) at least eight species related to this genus, including *H. tarsalis* Ashm, parasitize eggs of Bruchids.

The family *Eulophidae* includes almost twenty parasites of bruchids, which often present difficulties for taxonomists and justify Ferrière's expression 'the burden of the old species'. Indeed in the first half of the nineteenth century, many Eulophids were classified in a very superficial manner as *Entedon* and *Tetrastichus.* When they hatched from the pods of *Leguminosae* in which bruchids, *Apion* and Microlepidoptera could live together, it was decided, a little by guesswork, that the Chalcid attacked one of these insects, or two of them, even all three. So, according to Della Torre, *Entedon seminarius* Ratz, is the same as *Tetrastichus aethiops* (Zett.) and therefore would parasitize *Bruchus pisorum* L., while Ratzeburg found it in *Bruchus spartii* Eric (sic). But according to Domenichini (1965) *T. aethiops* attacks only Cynipids. This last author, however, believes that *E. seminarius* could be the same as *Tetrastichus spartii* (Ratz) which attacks *Curculio spartii* (sic) if we are to believe Ratzeburg, whereas Domenichini claims that this species is a hyperparasite of bruchids and of *Bruchophagus* (its real hosts being Pteromalids)!

Among the other *Tetrastichus* mentioned by De Luca neither *T. coerulescens* Ashm., nor *T. claviger* Thoms. can be included in his lists. *T. bruchophagi*

Gah. was imported from France to the United States in order to fight *Bruchus brachialis* Fähr.

To find out the real identity of *Eulophus* described more than a century ago and the 'types' of which are no longer available is as tricky a job as to pin down the exact classification of *Entedon* and *Tetrastichus*.

Although only one *Horismenus* (*Helcopeltomorpha*) can be found in Europe, in the New World this genus is well represented. Three species strictly dependent on the bruchids have been found. As already stated, one of them, *H. depressus* Gah. is one of the most common parasites of *A. obtectus* in Mexico. In the list given by De Luca, *H. popenoei* (Ashm.) should be crossed out as it is just another name for *H. missouriensis* (Ashm.). *H. bruchophagus* Burks, which hatches from *Mimosestes sallaei* (Sharp) and was first described in 1971 should on the other hand be added.

Pediobus bruchicida (Rond) is not apparently a parasite of bruchids, but a Lepidopterous hyperparasite (Boucek, 1974).

In the family *Eupelmidae* there are two genera, the members of which, as far as known, attack only bruchids. These are *Bruchocida** and *Charitopodinus*. However the largest number of parasites of bruchids can be found in the genus *Eupelmus* with five lariophagous species in the new world and two or three in the old one. In North America, *E. amicus* Girlt. not only attacks five species of bruchids, but also parasitizes 7 Lepidoptera 4 Coleoptera excluding the bruchids, and 2 Hymenoptera. As for *E. cyaniceps* Ashm. it has been found in six species of *Bruchidae* belonging to four genera *Acanthoscelides, Amblycerus, Gibbobruchus* and *Mimosestes*, 7 Lepidoptera, 11 Coleoptera, excluding bruchids, and 2 Ichneumonids. The range of hosts is even wider with *Eupelmella vesicularis* (Ratz): there is only one bruchid (*Bruchus brachialis*) but 11 Lepidoptera, 7 Coleoptera excluding the bruchid, and 25 Hymenoptera, Symphyta or Terebrantia, including 4 Chalcids. In De Luca's list *Eupelmus bifasciatus* Först and *E. inyoensis* Girlt. should be cut out.

In the family *Eurytomidae* one finds a return to phytophagy. Almost all the very numerous species belonging to the genus *Harmolita* Mots. (*Tetramesa* Walk.) pass their developmental stages in the stubs of a Graminacae. The transition from a parasitic way of life to phytophagy can be seen in some *Eurytoma* whose larva, after having fed on that of a *Harmolita*, completes its growth by feeding on sap. (*Eurytoma parva* Phillips and *Harmolita tritici* (Fitch).

The genus (or sub-genus) *Bruchophagus* Ashm. is composed of Eurytomids which being morphologically very close to *Eurytoma* are also phytophagous and grow, with very few exceptions, in the seeds of Leguminosae.** Yet even to-day one can find *Bruchophagus* classified in the literature as parasites of bruchids.

**Bruchocida batataephila* Riab. a parasite of *Cylas puncticollis* (Col. *Curculionidae*) is an *Eupelmus* (Boucek, 1976).

**Ashmead who had named this genus *Bruchophagus* thinking that its members attacked bruchids would never admit throughout his life that this was not the case. When people proved that they were phytophagous through experiments, he retorted that their observations lacked rigor.

On the contrary all the species of *Chryseidea* parasitize bruchids according to Burks (1971).

Among *Eurytoma* cited by De Luca as parasites of bruchids, to be on the safe side one should only retain the following: *E. bruchophagi* Blanch., *E. tylodermatis* Ashm., and *E. obtusa* Bugb. described in 1967 and probably introduced from Europe in the U.S.A. with its host *Bruchus brachialis* Fähr.

Systole albipennis Wlk. does not parasitize bruchids but grows on Coriander seeds.

There are a large number of Pteromalids which attack bruchids but the situation varies. *Anisopteromalus calandrae* (How.) which is synonymous with *Neocatolaccus mamezophagus* Ish. and Nag, *Aplastomorpha pratti* (Crawf.) and *Bruchobius medius* Ms., mainly parasitize Coleoptera living in stored food. As a result, they attack bruchids which can lay eggs in seeds extracted from their pods. The same is true of *Lariophagus distinguendus* (Först) and *Chaetospila elegans* Westw, whereas *L. texanus* Crawf. is more a field species.

Dibrachys cavus (Wlk.), extremely polyphagous, has been found in more than 200 hosts ranging from Forficulidae to Neuroptera, its main victims being Lepidoptera, Coleoptera and their parasites (Hymenoptera or Diptera). Yet, so far, it has only been found in *Bruchus brachialis* Färh.

For more than eighty years the taxon *Dinarmus* established by Thomson in 1878 was incorrectly interpreted by all the authors. As a result Mayr created the genus *Sphaerakis* in Europe and Ashmead the genus *Bruchobius* in America when these species should really have been included among the *Dinarmus*. All the species of *Dinarmus* (=*Sphaerakis*/*Bruchobius*) are thought to be parasites depending on bruchids. The two best known species are *D. acutus* Thoms and *D. basalis* (Rond.) otherwise known as *D. laticeps* (Ashm.) according to Boucek (1974). As for the *Dinarmus* auct. nec Thoms, they must be divided up into different genera created by Delucchi (1956). In many parts of the world these taxonomic changes are still not known. As far as the biology of *D. basalis* is concerned. Ferreira Lima (1942) made a major mistake when he claimed that the female of this species lays its eggs in those of the bruchid as soon as the embryo has been formed. Chatterji did not deny this assertion but Cheema and Misra definitely did.

The genus *Oedaule* Wtrst. which is very similar to *Dinarmus* is without doubt composed of species which are closely dependent on bruchids. The members of this genus can be found in large numbers. In the South of France, as we have seen, hundreds of *O. italica* Ms. fly around the pods of *Calytome spinosa* infested by *Bruchidius lividimanus* (01). *O. altifrons* (Wlk.) should be added to the list established by De Luca. On the other hand, *Habrocytus sequester* (Wlk.), *H. piercei* Crawf. *Pachylarthrus breviventris* Först (?), *Pseudocatolaccus bruchocida* Risb. should be suppressed. All these species are now classified under different genera and none of them parasitizes the bruchids.

Finally, on the American continent can be found the genus *Zatropis*, two species of which, *Z. incertus* (Ashm) and *Z. orontas* (*Z. bruchivorus* Ashm.) have been obtained from various *Acanthoscelides* and from *Bruchus brachialis* Fähr.

The *Torymidae* play only a secondary role in the pressure exerted by the Chalcids on the *Bruchidae*. It is probable that *Liodontomerus terebrator* Ms. which is a parasite of bruchids in the United States also attacks *Bruchophagus* (and not bruchids) in the rest of the world. On the other hand, *Antistrophoplex acithopoensi* (Risb) and *A. leguminosae* (Risb) classified under the generic name of *Senegalella* Risb parasitize *Bruchidius* (Boucek, 1976).

The family *Trichogrammatidae* includes a genus, *Uscana* Girlt., supposed to have no hosts other than the eggs of bruchids. We have had the opportunity of studying several *Uscana* parasitizing *Bruchus affinis* Fröl. in the South of France, *Callosobruchus maculatus* (L.) in Sudan (Mali), *C. chinensis* (L.) in the same country and *Bruchidius fasciatus* (01) in the Paris region. One notices that the number and disposition of sensillae found on the antennae of females as well as males remain the same for a given host. So one may then wonder whether each species of *Uscana* does not correspond to a species of bruchid and be surprised that in the USSR, *Uscana senex* Grese breeds from the eggs of three *Bruchus* and *Bruchidius*. It is true that Soviet authors do not give any precise information about sensillae to which they seem to pay little attention. In the United States, *U. semifumipennis* Girlt. is given what seem to be an excessive number of hosts: *Acanthoscelides alboscutellatus* (Horn), *Althaeus hibisci* (01), *Callosobruchus maculatus* (F), *Stator limbatus* (Horn). Peck only mentioned *C. maculatus* and maybe he was right. Whatever the truth of the matter, the number of *Uscana* classified in the world at present would reach thirteen.

The last edition of the catalogue of Hymenoptera in North America (united States and Canada) provides the most recent and comprehensive survey of *Bruchidae* and their parasites in a well investigated area. We learn from these data that 9 *Bruchidae* species are each attacked by only one species of Hymenoptera while *Bruchus brachialis* Färh can be the victim of at least ten parasites (9 Chalcididae and one Braconidae) (see Table 1).* It should not be inferred, that *B. brachialis* is, more than the other members of its family, a choice host for parasites. Actually, *B. brachialis* is much more detrimental for the Americans than the *Bruchidae* related to food legumes. Inasmuch as it attacks vetches, that species has been subjected to very elaborate inquiries and investigations, like the *Bruchophagus* which destroy the seeds of fodder crops, which quite naturally led to the discovery of many parasites. Their large number is a mere consequence of economic concerns. When looking for

Table 1.

Number of Bruchus species	9	10	–	1	2	1	1	2	–	1
attacked by 1, 2 . . . 10 parasite species	1	2	3	4	5	6	7	8	9	10

*The hosts cited by Peck (1963) and which are not mentioned in the catalogue of Hymenoptera in North America, have not been taken into account, except *Habrocytus piercei* Crawf.

a host (or ovipositing) many parasitic Hymenoptera seem to respond more to a given environment than to the systematic position of the host. The consequence is that a *Chalcididae* such as *Monodontomerus aereus* Wlk. attacks *Porthetria dispar* a leaf eating Lepidoptera, as well as larvae of Diptera or Ichneumonidae which would already inhabit the Lymantriidae's pupae. Therefore *M. aereus* is, as the case may be, a primary or a secundary parasite. For this extremely widespread phenomenon, which is due to a parasitic specificity that could be said to be nonexistent, the word 'metaparasitism' has been coined, but unfortunately it is little used. In fact, metaparasitism is quite different from constant hyperparasitism, which can be observed mainly in *Chalcididae.*

The presence of a parasite in a well defined site (or looking for that site) is often the consequence of a phytotropism, or of a tendency to stay in a particular plant layer (herbaceous, shrub, tree, etc. .). There are consequent absurd situations. In Java, some *Chalcididae* of the genus *Pleurotropis,* parasites of Hispids whose larvae live in the parenchyma of coconut palm leaves, ignore the Hispids which mine the other local monocotyledones, although they can breed on these in the laboratory. On the other hand, in nature, they attack the larvae of a Buprestid which also bore into coconut palms, though they are unfit for the development of *Pleurotropis.*

In such a case, there is no more host specificity, but a biotope specificity (the coconut palm frond), and a microbiotope specificity (the gallery bored in the leaf). There is nothing exceptional in that phenomenon which can be observed notably in many Hymenoptera which find their hosts without discrimination among Diptera, Lepidoptera and Coleoptera because their larvae have the common characteristic of living hidden either in a hollow stalk, or in a pod, or in a capsule, etc. . . Thus *Bracon variator* Nees successfully lays its eggs on *Myelois* caterpillars, on the larvae of weevils of the genus *Larinus,* as well as on the maggots of *Tephritidae* living in the capsules of *Carduaceae.* Their behaviour is similar to that of metaparasites. In that respect, it should be remarked that man divides time and space into units, for example, Forest, Tree, Leaf, Gall, which are also entities, and the fruits of his reason, but which do not exist as such for the insect. The insect's world is a continuum made up of changing associations of stimuli, one or several of which can prevail over the others, but which are beyond our perception. As a result, there is an extreme diversity in the choice of hosts: a near-lacking specificity in *Dibrachys cavus* (Foerst.), or the uttermost one in the Encyrtid *Prospaltella perniciosi* Tow., which has always been found in the world only on the San-José Scale *Quadraspidiotus perniciosus* (Comst.).

Apart from the nine species of *Braconidae,* the *Dinarmus,* the *Habrolepoidea,* the *Chryseidea,* the *Uscana,* the three *Horismenus,* the two *Eupelmus,* one *Eurytoma* and one *Zatropis,* which exclusively attack *Bruchidae,* the other 13 Chalcididae detrimental to the latter in the United States also attack weevils. Let us set aside *Anisopteromalus calandrae* How., *Lariophagus distinguendus* (Foerst.) and *Chaetospila elegans* Westw., which develop on *Sitophilus* and *Cauphilus* only because they belong to the vast group of foodstuff insects just like the *Callosobruchus,* and then it can be noted that all of the remaining species develop, in the fields at the expense of *Anthonomus*, and especially of *Anthonomus grandis* Boh., which is well-known owing to its damages on the cotton-plant capsules. The way of life of the *Anthonomus* accounts for their being the victims of bruchids' parasites. In addition, *Bruchidae* and *Curculionidae* are closely related. But above all, the following point should be emphasized: it is thanks to the immense efforts made in the South of the United States to search for the parasites of the Cotton Boll Weevil and their cohosts that six *Chalcididae* detrimental to bruchids were discovered, an example that to be followed.

REFERENCES

Boucek, Z., 1974. On the Chalcidoidea (Hymenoptera) described by C. Rondani. Redia 55: 241–285.

Boucek, Z., 1976. Changes in the classification of some African Chalcidoidea (Hymenoptera). J. ent. Soc. Sth. Afr. 39: 345–355.

Burks, B.D., 1971. A synopsis of the genera of the family *Eurytomidae*. Trans. Am. ent. Soc. 97: 1–89.

Burks, B.D., 1979. *vide Krombein et al.*

Cheema, P.S. & Misra, J.N., 1962. Mode of oviposition of *Bruchobius laticeps* Ashm (Hymenoptera: Chalcididae), a parasite on *Callosobruchus* (sic) sp. Curr. Sci. 31: 21–22.

Delucchi, V., 1956. Beiträge zur Kenntnis der Pteromaliden. Zeit. f. ang. Eng. 39: 229–257.

Domenichini, G., 1965. I. *tetrastichini* (Hymenoptera *Eulophidae*) paleartici ed i loro ospiti. Bol. Zool. agr. e Bachicol. (2° Ser.) 6: 61–203.

Ferreira Lima, A.D., 1942. Mais un inimigo do *Callosobruchus chinensis* Bol. Soc. Bras. Agr. 5: 441–443.

Krombein, K.V., Hurd, P.D. Jr., Smith, D.R. & Burks, B.D., 1979. Catalogue of Hymenoptera in America North of Mexico, 3 vols.: 1–2735.

De Luca, Y., 1965. Catalogue des Métazoaires parasites et prédateurs de Bruchides (Coleoptera). J. Stored Prod. Res. 1: 51–98.

Peck, O., 1963. A catalogue of the nearctic Chalcidoidea (Insecta, Hymenoptera). Canad. Ent., Suppl. 30: 1–1902.

Pierce, D.W., Cushman, R.A., Hood, C.E. & Hunter, W.D., 1912. The insect enemies of the Cotton Bool Weevil. U.S. Dept. of Agr., Bur. of Entom., Bull. No. 100: 1–99.

Taylor, R.A., 1937. The biological control of an insect in Fidji. London. Imp. Inst. Ent.: 1–239.

Author's address:
Muséum National d'Histoire Naturelle
45 rue de Buffon
75005 Paris
France

CONCLUSIONS AND RECOMMENDATIONS

CONCLUSIONS

The importance of legumes as food and fodder is now fully recognized, and this provided the basis for the work of the International Symposium on the Ecology of Legume Bruchids, organized in Tours by the Institut de Biocénotique Expérimentale des Agrosystèmes (IBEAS), of the French National Centre of Scientific Research (CNRS) and by the Université François Rabelais, in collaboration with the International Organization for Biological Control (IOBC) and the Food and Agriculture Organization of the United Nations (FAO).

The meeting stressed the vital role of legumes in reducing the protein shortage and thus improving the food and health situation of hundreds of millions of people, and proceeded to define how the availability of these resources can be increased.

The bruchids which attack leguminous crops frequently cause disastrous losses. Severe or total loss is for instance often recorded on beans (*Phaseolus vulgaris*) and cowpeas (*Vigna unguiculata*) causing serious temporary shortages especially at the end of the storage period (e.g. in Rwanda and Burundi).

In addition to attacking widely cultivated pulses, bruchids prevent the establishment of new or neglected species which are now recognized to be important and which could considerably improve the food situation in the poorest regions. Due to lack of systematic research, only 20 of the 90 species of pulses consumed by man have been studied in relation to bruchid losses.

Chemical treatment of stored pulses can reduce contamination in appropriately equipped silos or stores. However, these techniques are difficult to apply under rural storage conditions and cannot be used at the village and farmer levels. It is therefore of paramount importance to cut off infestation at its sources in the field. This is possible on the basis of an adequate knowledge of the biology of bruchids in the field and of the conditions under which crops and harvested pulses are infested (including the out-door drying period).

The following recommendations are accordingly addressed to FAO and other international and national organizations involved in technical aid, and in agricultural biological research, in the hope that research leading to a reduction in bruchid attack on legumes will make its contribution to closing the protein group.

Since FAO is in a position to play a decisive role in such an endeavour, the meeting called in particular on this organization to assist in research coordination and in the dissemination of information on bruchids and their control. Action along these lines should encourage national organizations to increase their efforts in three priority areas:

1. Surveys of actual and potential bruchid pests.
2. Ecology of bruchids on crops and on wild populations of pulses consumed by man.
3. Surveys of traditional methods of storage and bruchid control for harvested pulses.

RECOMMENDATIONS

These conclusions lead to the following recommendations.

Establishment of an international task force

The establishment of a task force dealing with harmful bruchids should be considered by IOBC and FAO. This group should include experts on the bruchids (taxonomists, ecologists, biochemists, physiologists, ethologists) as well as food legume experts (taxonomists, biochemists, physiologists and breeders); it should further include entomologists specialized in crop protection. The terms of reference of the group would primarily be to study the problems of bruchid control in the most severely affected areas. The group would have a short-term medium-term programme and could thus advise FAO on any action it could support at the national and international levels.

1. SHORT-TERM PROGRAMME

1.1 Establishment of a centre for the collection and distribution of data on actual and potential bruchid pests.

1.2 Collection and distribution of data on control, including methods used in regions of traditional pulse cultivation.

1.3 Study of ways to increase the working capacity of institutes and researchers involved in ecological studies of bruchids and their control and in the improvement of traditional methods; provision of easy access to varietal collections at legume research centres.

1.4 Study of the possibilities for taxonomic research on bruchids and for dissemination of results.

1.5 Provision of training opportunities for entomologists concerned with the protection of pulses, in institutes studying bruchid ecology.

2. MEDIUM-TERM PROGRAMME

2.1 Establishment of modern means of distribution for data on bruchid taxonomy.
2.2 Organization of training in the taxonomy of bruchids and their parasites.
2.3 Establishment of a programme to promote the ecological education of entomologists concerned with bruchid pests.
2.4 Study of ways of promoting research on harmful bruchids and their parasites in the centres of diversification of edible pulses and in characteristic regions where they are cultivated.
2.5 Establishment of field trials on the ecology of bruchid pests under natural conditions, particularly in arid regions and in the high-altitude equatorial zone.
2.6 Establishment of a documentation centre for the ecology of bruchid pests allow the coordination of this study.

Paris, 8 July, 1980

MIX
Papier aus verantwortungsvollen Quellen
Paper from responsible sources
FSC® C105338

If you have any concerns about our products,
you can contact us on
ProductSafety@springernature.com

In case Publisher is established outside the EU,
the EU authorized representative is:
Springer Nature Customer Service Center GmbH
Europaplatz 3, 69115 Heidelberg, Germany

Printed by Libri Plureos GmbH
in Hamburg, Germany